PERPETUAL MOTIC

-INDEX

---

INTRODUCTION P. 11
THEORY P. 15
1st PERIODIC TABLE O' PMMS P. 26

Early Failures P. 29
Grav-Buoy P. 32
Fluid Leverage, P. 41
Curving Rail, P. 44
Motive Mass Machine, P. 45
Repeating Leverage, P. 58
Modular Trough Leverage P. 74
Tilt Motor, P. 75
Not-If-But-When Machines, P. 76
First Fully-Proven PMM, P. 82
Escher Machine, P. 84
Coquette, P. 86
Magnet Concepts, P. 90
Bezel Weight Device, P. 92
Gravity Motor, P. 93
Spiral Wheel, P. 95
Pendulums, P. 98
Spinning Tops, P. 101
Apollo Device, p. 103

OTHER DEVICES
ODDITIES, P. 105, APERRATURES, P. 110

HOW TO BUILD A MODEL, P. 115
ADDITIONAL THEORIES, P. 123
APPENDIXES P. 153

BIO P. 170

*Nathan Coppedge*

**PERPETUAL MOTION MACHINE DESIGNS & THEORY**

## THE LEGENDARY POEM

While they were floundering
I was pondering:
No more wandering through the
dark tunnels of grim determination
For no---it is time to grow in a thou-
sand folded folds
For which we need an infinite fuel!


---NATHAN COPPEDGE

*Nathan Coppedge*

**All content © 2006, '07, '08, '09, '10, '11, '12, '13, '14, '15, '16, '18 Nathan Coppedge**

# NATHAN COPPEDGE'S

# PERPETUAL MOTION MACHINE

# DESIGNS & THEORY

## Updated Edition

Nathan Coppedge

*Nathan Coppedge*

# PERPETUAL MOTION MACHINE DESIGNS & THEORY

## PREFACE

When I was a child every night I would have a dream. In the dream I would try to fall asleep in a children's clubhouse that I built in the yard with my family in real life. Except, the entrance to the clubhouse was packed with all manner of scrap metal.

Later, when I was 33 I had the same dream only the clubhouse, once dark and dingy, had been transformed into a luxurious house. When I looked at the scrap metal, it changed into something that looked mechanical, and was made of gold and brass. Perhaps it even moved? With tunnel vision, I could see the mechanical thing in the future—-in the inner sanctum of a kind of temple.

I hardly ever remembered the dream from my childhood until I turned 33 and the meaning became more clear.

The dream seems to be symbolic for the quest for perpetual motion. If you will speak to a psychologist they will tell you that dreams about going to sleep are especially significant. They are dreams that deal with the deepest occupations of the mind, thoughts from the well of the unconscious...

I often think my preoccupation could be more important than scientists assume...

*Nathan Coppedge*

# INTRODUCTION

*Nathan Coppedge*

# PERPETUAL MOTION MACHINE DESIGNS & THEORY

## -INTRODUCTION-

Perpetual motion is a field that has since the time of Leonardo Da Vinci, undergone significant scrutiny. Johann Bessler, one of the few designers who claimed to have actually built one, has been widely regarded as a fraud, although some reports say that his device was working after being shut in a room for three months. Although Isaac Newton did not witness the Bessler Wheel first-hand, it is rumored that he would have been critical. Mysteries like this obfuscate the underlying difficulties in constructing---or even, designing, such an invention.

In my own work, I have developed a practical eye towards the extreme difficulty in designing and proving them.. I have undergone a process like a supercomputer, (1) proving originality, (2) proving variety, (3) improving designs, and (4) seeking proof in experiment.

If there was a fault in this process, it was the difficulty of construction. It was years before I thought I found any evidence, even then only in partial experiments. Around Nov 2013 the evidence finally became highly compelling.

Because experimentation was so difficult---I was not the greatest builder, only a very expert designer---I focused on formulating designs which others could later prove mathematically. I also did my best to formulate my own proofs, although these were often not completely mathematical.

The major purpose of this book is to serve as a record of significant perpetual motion design principles. The approach addresses the problem of proportionality---the fundamental problem----in order to turn physicists' assumptions on their head. I am referring to principles such as the equivalence of velocity and mass in the momentum equation, and the principle that balancing scales can equilibriize.

My most recent designs, such as the Fully-Proven Perpetual Motion 1, and the Not-If-But-When Machines are designed to produce constant momentum, and overcome the cycle-return problem through coincidental proportionality.

I rely on strategic principles, some of them so elegant as to be invisible except from the correct theoretical standpoint.

This book is designed around proving to the creative mind that perpetual motion is possible. If there is something missing it is conventional physics. I will add that in if I have the opportunity...

# THEORIES

*Nathan Coppedge*

# 10 PROOFS

1. Friction does not eliminate motion where motion is permitted.
2. Reactions are possible in a circle, as shown by dominoes. Wheels can turn.
3. Dominoes can chain-react using higher and higher altitudes. Energy can be created.
4. Dominoes can accelerate, so friction does not stop everything.
5. In principle, equilibrium is enough to overcome proportionality problems.
6. Imbalance can overcome friction.
7. Equilibrium and imbalance can exist simultaneously through mass- leverage ratios.
8. With unbalance and a principle of momentum, there is no need to lose altitude over time.
9. All else considered, natural momentum with no net loss of altitude = perpetual motion.
10. With natural momentum and upward and downward motion, potential for return.

*Nathan Coppedge*

# PERPETUAL MOTION MACHINE DESIGNS & THEORY

## "VOLITIONAL MECHANICS"

FIRST PRINCIPLE: Continuous Motion:
However unbalanced the parts, there must be a method of continuum, a means such as ramps and differences in force, to allow motion to continue. This is the most central principle, but by itself it does not yield much technicalism.

SECOND PRINCIPLE: Unity
In order to meet criteria of unity in a simple device, all parts must interact via equilibrium, such as over an axle, or through circular motion. If not, then this suggests use and disuse of parts, which is still a function of equilibrium. Use and disuse can also be an effective principle, if there is no change-of-altitude problem.

THIRD PRINCIPLE: Over-Unity
Over-unity is essentially the principle of motion in spite of equilibrium. This can be done through directed energy (continuous slope, as in the Tilt Motor), differential energy (mass-space differences, as in Repeat Lever Type 2), or an unbalanced principle (methods of "cheating," as in the Trough-type devices).

FOURTH PRINCIPLE: Volitional Energy
is represented by the Volit (pictured), representing an effective cycle. Devices tend to be more functional not only with a good principle, but with a large number of moving parts in comparison to fixed and dual-axial parts. To my knowledge this has borne out with the exception of "cheating" principles, in which fixed units may actually have an advantage.

Ultimately I decided dual-axialism was a more important disadvantage than fixed parts in the right design. But, instructively, both may be considered bad. Volitional Energy is expressed as:

= Active Units / Dual-Axial Units

FIFTH PRINCIPLE: Volitional Equilibr. (Ve) Generally, a large number of modules is good when it works, and a large number of channels or branches in the movement is bad. Not all devices use modules, so this value is often 1. It is expressed as:
Ve = (mod. U /
(branches / cycle) / (subcycles / cycle) )

SIXTH PRINCIPLE: Volitional Eff. (VE) Overall efficiency is estimated as an equation between volitional energy and volitional equilibrium. For example, if volitional energy (also called volition) is 2, and equilibrium is 1, the result is 2. If the equilibrium is 4, the result is 0.5. If the equilibrium is 0.5, the result is 4. Higher numbers are better. You will find for most viable principles, the number is hardly ever above 2. '1' represents unity, hypothetically, and the number must be greater than 1 to equal perpetual motion.

$$VE = \frac{\text{☿}}{Ve}$$

# PERPETUAL MOTION MACHINE DESIGNS & THEORY

**Three Types of Perpetual Energy**

Continuous Mass: Applying weight without motion

Continuous Velocity: Photons that have momentum without mass

Continuous Momentum: Machines that are unbalanced at every point in a cycle


**Principles of Volitional Energy**

Principle of Volitional Energy: *Mass that can by some means move on its own, and repeat this motion, has a form of momentum.*

Principle of Volitional Equilibrium: *Force within a cycle that does not divide is singular, whether or not it encounters resistance.*

Principle of Volitional Efficiency: *When the division of force is less than the implication of momentum, the function of a device scales to design principles, rather than mere re-enactment of basic laws understood in basic ways.*

There is freedom to create a machine whenever such a machine is possible to build. The process is a posteriori and not a-priori: it is found after the fact. As I will show later on, perpetual motion is a special case of specific relations of laws, laws which do not contradict the most reasonable perspective on conventional physics, in all conventional cases, and certainly would not contradict all unconventional physics, in unconventional cases.

## Major Theory Principles

*CYCLES CAN RETURN TO THEIR STARTING POINT*

Consider a circle of dominoes. The dominoes can return to the starting point. The problem is re-setting the cycle. Similarly, a wheel can be made to rotate in a circle many times, thus repeating the cycle; the problem is creating motion in the first place. These examples prove that cycles can certainly return to their starting points. There is no physical law against recursive movement.

*PERPETUAL MOTION DOES NOT REQUIRE ENERGY*

Momentum is Possible Without Velocity, Therefore Momentum is Possible Without "Energy": Just as photons have momentum without mass according to the basic rule of Mass * Velocity = Momentum, in certain cases mass that is not moving still exerts force, and since the mass does not move, the force is not lost. However, that force, according to my rules, is understood as momentum without velocity. An example of momentum that exerts "force" without losing energy is the case of a rolling object tethered horizontally upon a slope, with the tether leading horizontally (say, through a split in the sloped surface) beyond the slope, to some other fixed object. It is clear that the tether remains taut if the rolling object is heavy or if the slope is steeply inclined, yet with some allowance for control, there is no reason to believe that the tautness is caused by an actual movement of the rolling weight. For example, consider that any initial movement could be easily repeated so long as slope remains for motion. But there is no necessity that movement occurs in the first place at all. Certainly it is not a requisite for making the cord, rope, or string taut. The point is, however, that a general principle of momentum without velocity is potentially foundational for perpetual motion, because most physicists seem to have been assuming that energy was necessary for perpetual motion. The fogginess of the basic equation in which velocity or momentum replace energy suggests that any one of the three---velocity, mass, or momentum---may be energy in itself. While this does not necessarily make sense in the strictest physics equations, it is a foregone conclusion that each of those three aspects mentioned *does* connote energy in some sense, that is, they could mean potential energy *if* we assume gravity, and make the radical stipulation that gravity is not always consumed. We do not at this stage have to posit the creation of energy---indeed, the creation of energy must be a special case---and it is already by the adoption of a special case that one individual principle, such as velocity, mass, or momentum, could become viable. According to the principles of volitional energy, fractionism implies that only one of a coherent set of principles may be necessary to simulate the entire conjunction, that is, if the other principles are also present at other points. The principles of spatial difference and fractionism underlie this principle, of momentum without velocity.

*SYSTEMS CAN GAIN ENERGY*

    A case is raised of dominoes set up in increasing heights. Some people suspect that this would not work. Taken to a far extreme, they're right. They expect that a domino won't knock down a building for instance. It won't. However, one domino can knock down two dominoes stacked up. Try it. It works. Further, the two dominoes can knock over three dominoes. And the three dominoes can knock down four dominoes. It can keep going ad infinitum so long as it is possible to stack up the dominoes. Some would say that this flies in the face of physics. I would say that it is a fine application of physics. Also, consider as a stipulation that the altitude rule is not what is applying here. Don't be fishy about this issue. Increasing heights does not mean increasing altitudes, inherently. The dominoes could be stacked at decreasing altitudes, and they would still operate. That is enough to prove the point. Or, the first domino could touch a stick that reaches up to the top of a stack that then causes a downward chain reaction. How the chain reaction occurs has little to do with altitude, except in relation to changes in the gravitational force. Certainly we wouldn't say that a change in the gravitational force is what causes a chain reaction. Instead we would say that 'inputted' force causes motion (momentum), and in this case it is evident that a small inputted force is sufficient to cause a mass reaction. One of the only principles preventing this from becoming a self-repeating cycle is the abstract principle that the cycle cannot be re-set. It is obvious that the system can gain energy.

Major Theory Applications
*Energy
>Power & Heat
*Mobility
>Linear Transit through Power
>Cyclical Transit through Cont. Motion
*Social Reform
>Energy Independence
>Ecology & Arcology
>Free Money
*Symbolism
>Immortality
>Industry
>Intelligence
>Fun


Minor Theory Applications
*Continuous Motion
*Toys and Novelties
*Frivolity
*Intellectual Concepts

**PERPETUAL MOTION MACHINE DESIGNS & THEORY**

Perpetual Motion Types

Rank of Types and Variations

Escher Machine: 1 ∞
Motive Mass: 1 ∞
1st Fully Proven: ∞ infinities
Tilt Motor: 5/8 or 5 ∞
Spiral Cone: 1 ∞
NIBW3: 2 ∞
Repeat Lever 2 M1: 2 ∞
NIBW4: 2 ∞
Pendulum 1: 2 ∞ or: unity.
Unity Balls: Unity.

*Nathan Coppedge*

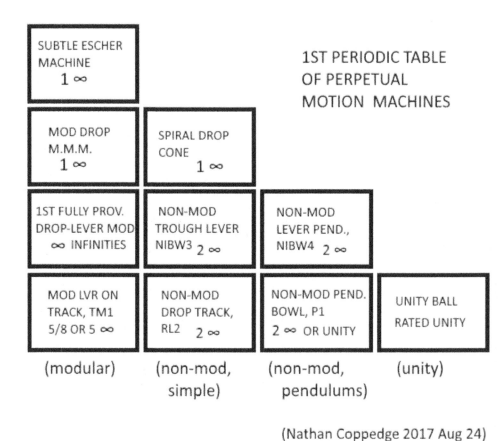

(Nathan Coppedge 2017 Aug 24)

**PERPETUAL MOTION MACHINE DESIGNS & THEORY**

*Nathan Coppedge*

**PERPETUAL MOTION MACHINE DESIGNS & THEORY**

# EARLY FAILURES

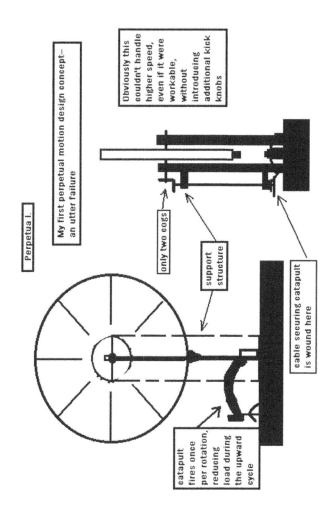

TYPE 1

*Nathan Coppedge*

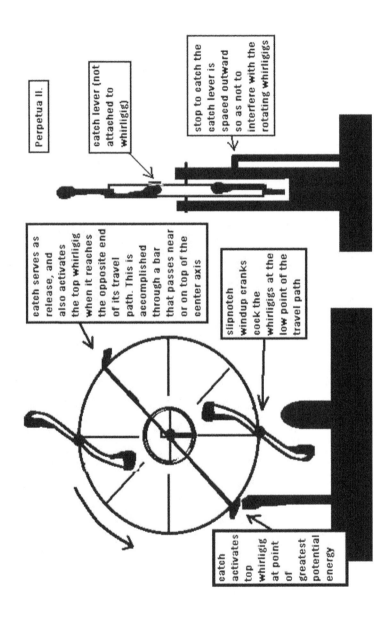

**EARLY CONCEPT 2**

**PERPETUAL MOTION MACHINE DESIGNS & THEORY**

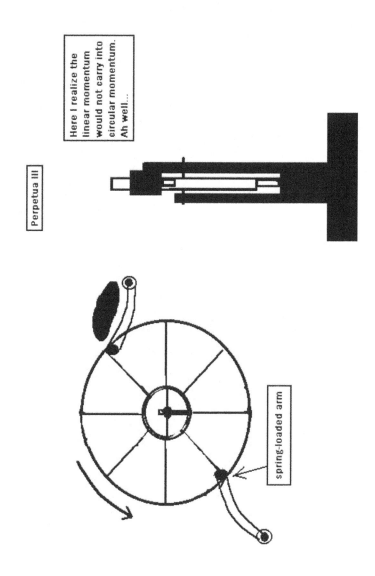

**EARLY CONCEPT 3**

## GRAV-BUOY ITERATION 1

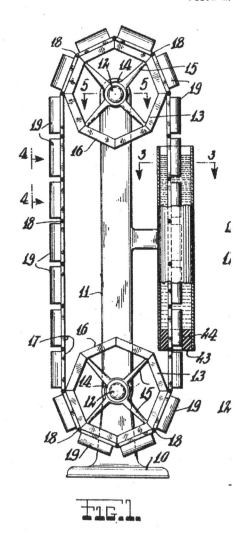

FIG. 1.

**FRANK TATAY'S DESIGN**

PERPETUAL MOTION MACHINE DESIGNS & THEORY

# GRAV-BUOY ITERATION 2
## COMPONENTS

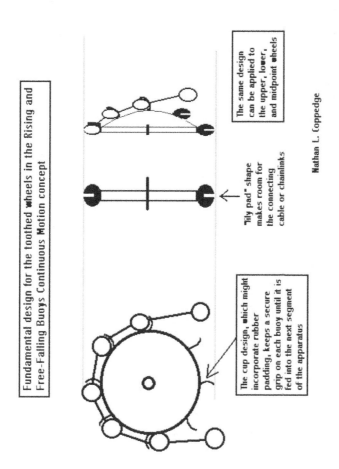

TOOTHED WHEEL FOR THE GRAV-BUOY 2

*Nathan Coppedge*

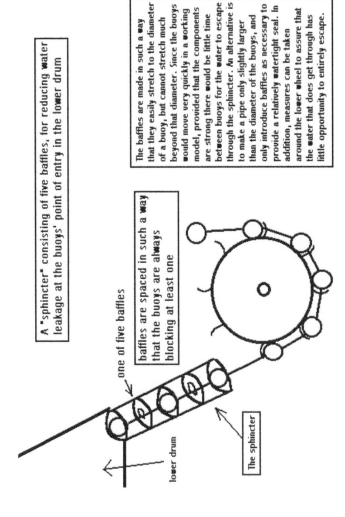

**SCHINCTER FOR THE GRAV-BUOY 2**

## GRAVITY-BUOYANCY DEMO

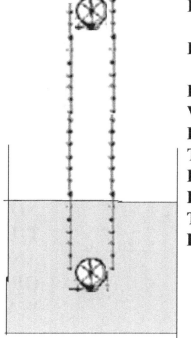

HYPOTHESIS 1

BASIS: FRANK T.

LOWERING THE WATER LEVEL IS THE BEST THEORY AT FIRST GLANCE HOWEVER, THIS PROVES EQUILIBRIUM

GRAV-BUOY DEMO 1

## HYPOTHESIS 2

## ALTERING THE ANGLE OF THE DEVICE IS THE EASIEST MOST LIKELY WAY TO EFFECT A CHANGE IN THE APPLICABILITY OF UNIVERSAL LAW

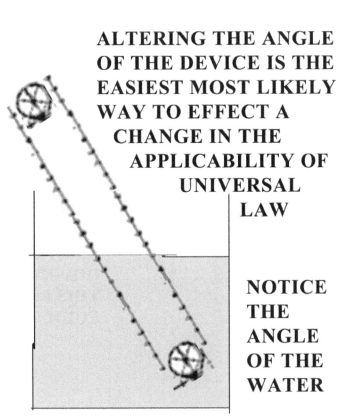

## NOTICE THE ANGLE OF THE WATER

GRAV-BUOY DEMO 2

## HYPOTHESIS 3

## A FURTHER PRINCIPLE WOULD MAKE SOME USE OF AIR PRESSURE HOWEVER, THE POSITION IS INITIALLY CONFOUNDED

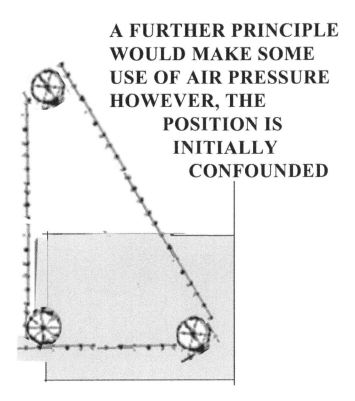

**GRAV-BUOY DEMO 3**

Nathan Coppedge

## HYPOTHESIS 4

## SOME SORT OF WATER TANK MIGHT BE RE-INTRODUCED, BUT THE PROBLEM IS WITH A BALANCE BETWEEN WEIGHT AND BUOYANCY, WITH DEVICIVE WATER PRESSURE AS AN INTERMEDIATE

**GRAV-BUOY DEMO 4**

**PERPETUAL MOTION MACHINE DESIGNS & THEORY**

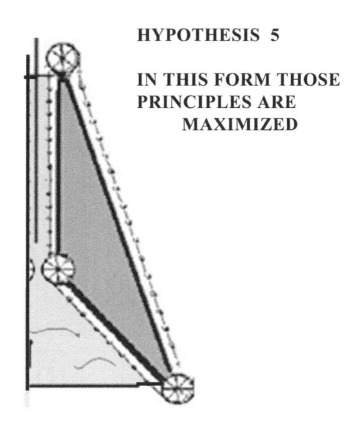

**HYPOTHESIS 5**

**IN THIS FORM THOSE PRINCIPLES ARE MAXIMIZED**

**GRAV-BUOY DEMO 5**

Nathan Coppedge

**Maximized efficiency**

In this variation the weight of the water is minimized, by distributing a narrow upper column into a wide lower column. Thus there is a strong gain in buoyancy in the upper column, while there is also much reduced entry resistance in the lower.

At the same time there is a risk that the gravitational assistance would be significantly reduced, due to the less than vertical angle of the outer sides

It may be noted that while increase in the height of the device relative to the size of the lower tank increases bottom pressure, proportionally more energy is generated, since additional rising buoys are added without much increase in water weight. Also, the angle of the free-falling buoys becomes more vertical, increasing the effective gravitational assistance

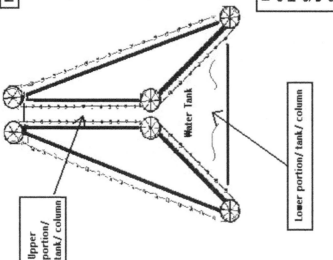

**FULL DESIGN FOR THE GRAV-BUOY TYPE / ITERATION 2 (SECTION)**

**PERPETUAL MOTION MACHINE DESIGNS & THEORY**

## FLUID LEVERAGE: THREE DESIGNS

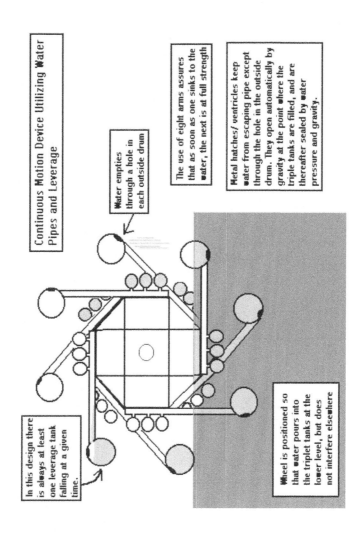

**FLUID LEVERAGE: LEVERAGE PIPES (FIRST DESIGN)**

*Nathan Coppedge*

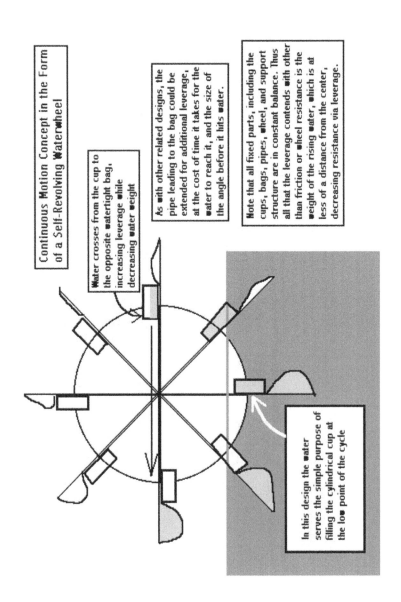

**FLUID LEVERAGE: WATERWHEEL
(SECOND DESIGN, ASSUMED MEDVL)**

**PERPETUAL MOTION MACHINE DESIGNS & THEORY**

FULLY BUOYANT WHEEL USING BUOYANT SLATS

sealed buoyant wheel, partially submerged

Here a buoyant wheel is used, constructed in a way where the slats of the wheel are buoyant as well; The wheel is partially submerged in water, so that there is no resistance at the top of the wheel; The principle is that the buoyancy is assymetric at every point; if so, it seems perpetual

buoyancy pushes up, favoring motion only vs. resistance

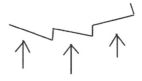

## OILSKIN, BUOYANT WHEEL (THIRD DESIGN)

Nathan Coppedge

## CURVING RAIL DEVICE

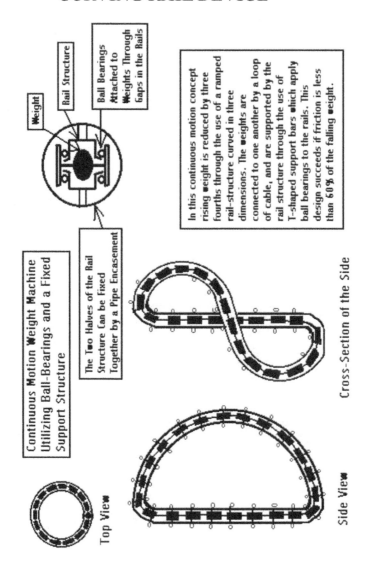

**CURVING RAIL DEVICE / WEIGHT MA-
CHINE (DESIGN 1/1 FOR THIS TYPE)**

PERPETUAL MOTION MACHINE DESIGNS & THEORY

## MOTIVE MASS MACHINE
### EXTENSIVE DESIGNS

Continuous Motion Concept Utilizing a Weighted See-Saw and Movable Difference Weight

Outer Curving Tube

This design takes advantage of the effect of applying minimal force to alternate ends of a weighted see-saw. Minimal weight differences can cause the whole structure to shift, creating greater force (mass times distance) than need be applied to the difference weight. I'm speculating that the large weights on either end can be used to operate catapults which fire an outer weight along a curving tube. If that outer weight is attached by a cord or cable through a slit in the tube, it might provide sufficient force to move the difference weight, creating a perpetual cycle.

The ball passing along the outer rail might just as easily be a wheel on a track, or a weight attached to a T-frame with ball bearings along a track. That might minimize resistance.

returned force

difference weight

weight

Detail of one application

On a moderate scale the device might be used to operate pneumatic pistons or crank up two ratcheted wheels. Multiple such devices might be combined in clever ways for a larger overall effect.

### SEESAW APPARATUS
### (DESIGN 1/13)

*Nathan Coppedge*

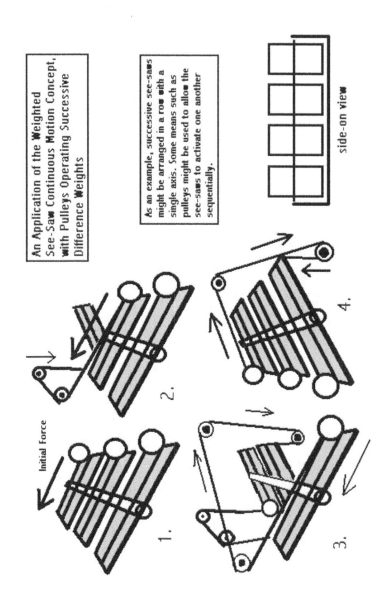

**HOW TO TRIGGER A CHAIN REACTION**
(BASIC RELATED CONCEPT)

# PERPETUAL MOTION MACHINE DESIGNS & THEORY

**An Application of the Weighted See-Saw with Difference Weight Continuous Motion Concept**

In this application two sets of weighted see-saws are used with a single outer tubed weight. The single outer weight is attached to the difference weights of both see-saws, so that slightly more than double pull is required, but provided that that pull can be generated by a single weight, two see-saws can be used more efficiently within a smaller space, leaving room for additional machinery. Doubling the number of machines active on a single catapult may permit relatively higher force applied to the outer tube weight.

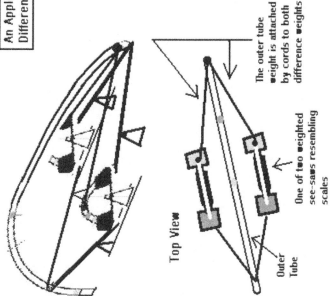

The outer tube weight is attached by cords to both difference weights

One of two weighted see-saws resembling scales

Top View

Outer Tube

**CATAPULTING ARCH CONCEPT
(CONCEPTUAL)**

Nathan Coppedge

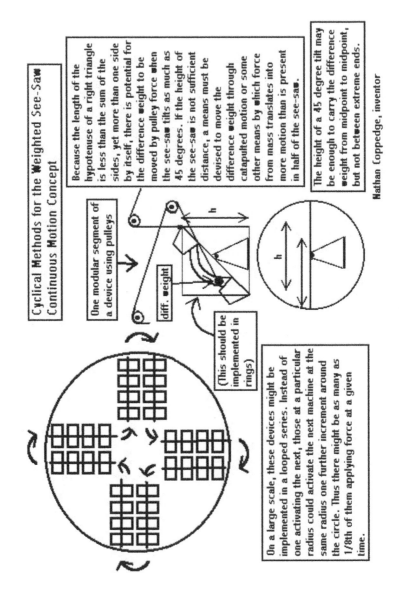

**MOTIVE-MASS CYCLICAL METHOD
SHOWING PROPORTIONAL ADVANTAGE**

# PERPETUAL MOTION MACHINE DESIGNS & THEORY

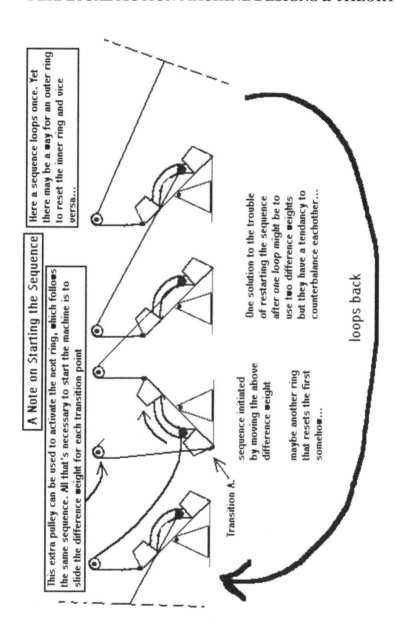

PULLEY CONCEPT FOR THE
MOTIVE-MASS MACHINE

# Nathan Coppedge

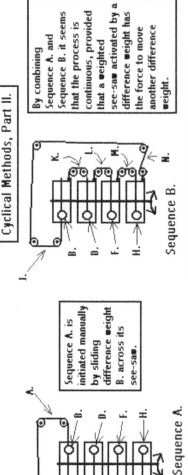

**MOTIVE MASS MACHINE CYCLICAL METHOD PROCESS DIAGRAM**

# PERPETUAL MOTION MACHINE DESIGNS & THEORY

A rolling wheel can be pulled some distance if a free-falling equal mass is connected to it by pulley. This is because the weight of the rolling wheel is supported by the ground, whereas the falling weight transfers much of its mass into pull. If we look carefully at the diagram at left we can see that counterweight A. is supported by the track B. when pulled, but that the entire mass of A. contributes to downward pull C. through pulley D., once it has been pulled across the track.

The result is that when the weight of counterweight A. plus the smaller upper portion of the track B. is significantly greater than the larger lower portion of track B., and an initial force is applied, then one such device has the potential to activate the next in succession.

Because of the balance of masses E. and F., a significant movement of mass is possible with the application of minimal difference. The leverage force applied through the center by the counterweight allows a return on the mass of the fixed weights.

**THEORY OF MOTIVE-MASS**

*Nathan Coppedge*

### Dual Motive Mass System

In this design one motive mass see-saw is mounted on top of a larger one, thus maximizing return on the single difference weight.

The design operates in three stages per cycle. (1) The difference weight is moved. (2) The first see-saw responds to the difference weight, applying force on the larger see-saw. (3) In response, the large see-saw also tilts.

Theoretically the movement of the larger see-saw would provide more than enough force to move another difference weight, or reset its own. Note that the entire system is essentially in the same position afterwards as it was at the beginning, meaning that there is no loss of potential energy.

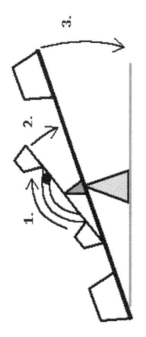

Application of this design in series would be similar to my most effective earlier implementation, the difference being that each unit would consist of a dual system, the difference weight pulled would be the single difference weight present here, and the larger see-saws would operate the pulleys. Note that in this design the full falling length of the larger see-saw has greater potential to move the difference weight the full distance, as it is proportionally greater than in previous designs.

**DUAL-SEESAW CONCEPT**

# PERPETUAL MOTION MACHINE DESIGNS & THEORY

## Dual Motive Mass System—Analysis of Principles

There are several things worth noting here in response to potential criticisms:

Potential criticism
"Only the difference weight contributes to the force of the larger see-saw, since the smaller see-saw is balanced"

Wrong. If the smaller see-saw can be tilted by the difference weight, the weight over time of the difference weight applied to the mass of the smaller see-saw is converted into an initial force acting from the smaller see saw to the larger one. If that force is sufficient to move the larger see-saw, we can see that the center of mass of the smaller see-saw begins to tilt. As it does so, the smaller see-saw begins to apply additional pressure on the large one by virtue of the fact that the leading weight is further along its side.

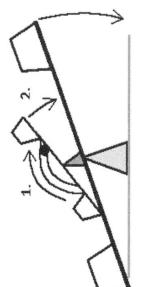

Note that the larger see-saw will begin to seek equality as soon as the left portion of the smaller see-saw moves upward. Thus the initial impact at 2. actually occurs while the larger weights are already in motion, creating a cumulative effect. That also explains how energy might be generated by the movement of the smaller see-saw: once the difference weight meets the midpoint, the see-saw is already seeking equality, so that the difference weight only adds to the effect.

**DUAL-SEESAW ANALYSIS**

*Nathan Coppedge*

**Dual Lever Motive Mass Machine—Initial Concept**

Here several different positions are depicted for the three weighted arms. The two lower arms are fixed at an acute angle from one another, and are hinged at the joining point. The upper weighted arm alternates, leaning on one of the lower weights and then the other, through use of a hinge fixed not to the acute arms, but to a separate support structure.

In principle this device works because the lower arms come into alternate distances from their axis. The majority of the upper weight is supported by the upper hinge, so that if force is provided to move it approximately 45 degrees, it then makes a return on force through the use of the acute arms, which may serve as levers.

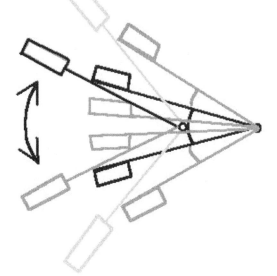

**DIFFERENCE-LEVER (DUAL LEVER) CONCEPT**

# PERPETUAL MOTION MACHINE DESIGNS & THEORY

## MOTIVE MASS ITERATION 2

**An Analysis of the Principles Operating in the Second Iteration of the Motive Mass See-Saw Concept**

This is a difference weight method using a triangular track which runs upward at D. and tilts upon application of weight at C.

We can note several things about the principles operating.

1. While the length of the upward track D. is greater than the distance that a weight at C. could pull through height B., if the difference weight is pulled strictly on a horizontal, for example if the weight has an axle that runs through it with cords attached on either side of the axle, then the weight need only be pulled a lesser distance of cord along the horizontal. That distance in this case is equal to A. Since A. is equal to B. the full weight applied at C. may be transferred into moving a weight sufficient distance to perpetuate the process.

Nathan L. Coppedge

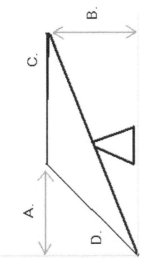

2. In this particular geometrical configuration, the length of the track at C. is actually sloped downward even when it is at its greatest height. This allows the difference weight to roll the remaining distance of the see-saw after it has reached the midpoint of the track.

## MOTIVE MASS IT2 THEORY

*Nathan Coppedge*

> An Application of the Second Iteration of the Motive Mass See-Saw Concept in a Vertical Series, Consisting of Two Cycles

An over-unity method consisting of three specially built see-saws with rolling difference weights, connected by pulleys.

**Cycle I**

The first cycle begins by manually moving difference weight A. towards pulley P. The force of the weight is transferred through see-saw B. to a cord attached on the right side of see-saw B. through pulley C., pulling difference weight D. As a result, force is transferred through see-saw E., across pulley F., pulling G. across H., transferring force through I. and J., pulling A. back across.

**Cycle II**

Force is then transferred through pulley K., moving D. back across, transferring force through L., moving G. back across. See-Saw H. then moves A. across for a third time, through pulley P. Since the fastening between O. and N. has been attached during the first cycle, difference weight A. is now sealed into the cycle as much as D. and G.

Nathan L. Coppedge

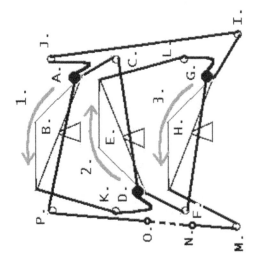

> Remember, this is theory, not fact. I haven't built this device.

**FINAL DESIGN FOR MOTIVE MASS (ITERATION 2)**

**56**

**PERPETUAL MOTION MACHINE DESIGNS & THEORY**

**CHAIN-REACTION TOWER CONCEPT**

### Chain Reaction Tower Perpetual Motion Concept

Difference lever (A.) is toggled, excentuated by stouter difference wedge (B.), operating a similar longer difference wedge (C) via a bracket; Process continues in the blockier lower structure by a domino principle; Cycle resets through operation at bracket (D.) and cross-beams at (E.); The three sets are designed to move in alternating directions, creating a chain reaction::

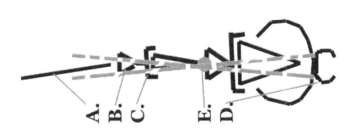

Nathan L. Coppedge

MOTIVE-MASS CHAIN-REACTION TOWER

Nathan Coppedge

## REPEATING LEVERAGE APPARATUS

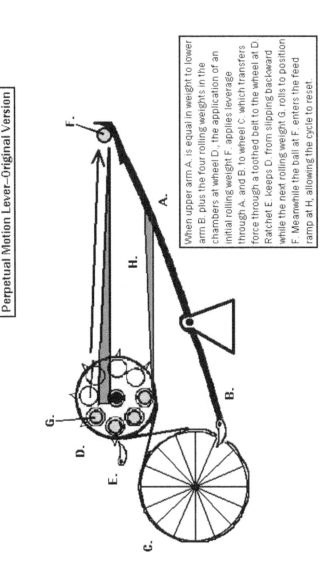

DESIGN / ITERATION 1

# PERPETUAL MOTION MACHINE DESIGNS & THEORY

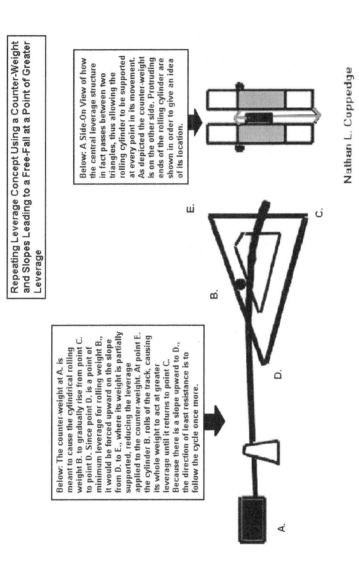

Repeating Leverage Concept Using a Counter-Weight and Slopes Leading to a Free-Fall at a Point of Greater Leverage

Below: A Side-On View of how the central leverage structure in fact passes between two triangles, thus allowing the rolling cylinder to be supported at every point in its movement. As depicted the counter-weight is on the other side. Protruding ends of the rolling cylinder are shown in order to give an idea of its location.

Below: The counter-weight at A. is meant to cause the cylindrical rolling weight B. to gradually rise from point C. to point D. Since point D. is a point of minimum leverage for rolling weight B., it would be forced upward on the slope from D. to E., where its weight is partially supported, reducing the leverage applied to the counter-weight. At point E. the cylinder B. rolls of the track, causing its whole weight to act at greater leverage until it returns to point C. Because there is a slope upward to D., the direction of least resistance is to follow the cycle once more.

Nathan L. Coppedge

**REPEATING LEVERAGE TYPE 2**

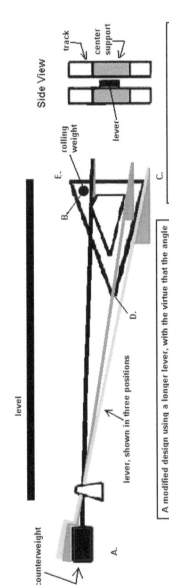

**TYPE 2, VARIATION 1**

**PERPETUAL MOTION MACHINE DESIGNS & THEORY**

Repeat Lever Type 3
Eucaleh Terrapin, Inventor

Lever passes in slot through guides for descending ball weight

A design using three ball weights and a double chambered structure lifted by the short end of a lever. According to the proportions as pictured the movement of a single weight at A. would be sufficient to lift both chambered weights, the topmost weight being incremented onto a gentle ramp at B. towards point A., the bottom-most weight being incremented into the upper chamber through a side ramp (not pictured), and the entire operation facilitated by a counterweight by pulley, equal to slightly less than one chambered weight (one chambered weight being constant)

**REPEAT LEVER TYPE/ ITERATION 3**

<u>61</u>

*Nathan Coppedge*

Repeat Lever Type 4   Eucaleh Terrapin, Inventor

A method in which two ball weights follow a variable path, the upward portion extending in a crescent followed closely by a mobile guide bar shaped to support one ball weight along the upward grade and attached to the longer end of a lever,

the short end cupped to take a descending wieght, resistance being minimized by the partial support of the rising weight by the fixed crescent track.

upward movement

downward movement

downward movement

**REPEATING LEVERAGE TYPE 4**

**62**

## PERPETUAL MOTION MACHINE DESIGNS & THEORY

### ARCHEMECHANICS: Trough Lever Device

Double-trough device in which a partial trough assists upward leverage via a fixed half-track:

TOP: SIDE VIEW of trough leverage device

The dynamic is meant to promote energy generation by movement of a fulcrum, since greater weight bears on the mobile track when the ball weight is not supported by the fixed track segment

RIGHT: section view of the mobile dynamic of this continuous motion device

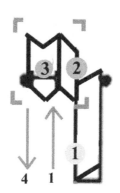

(c) 2011 Nathan Coppedge

### REPEAT LEVER TYPE / ITERATION 5:
BELIEVED TO BE FUNCTIONAL IN SOME FORM: MAKES USE OF GREATER WEIGHT BEARING ON MOBILE THAN FIXED APPARATUS, USES RAMPS TO DELIVER WEIGHT IN AND OUT OF APPARATUS

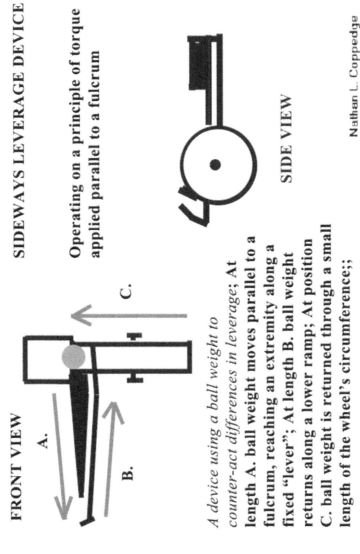

## SIDEWAYS LEVERAGE DEVICE

Operating on a principle of torque applied parallel to a fulcrum

*A device using a ball weight to counter-act differences in leverage; At length A. ball weight moves parallel to a fulcrum, reaching an extremity along a fixed "lever"; At length B. ball weight returns along a lower ramp; At position C. ball weight is returned through a small length of the wheel's circumference;;*

**SIDEWAYS LEVERAGE DEVICE: A FAILED CONCEPT (HAS BEEN TESTED)**

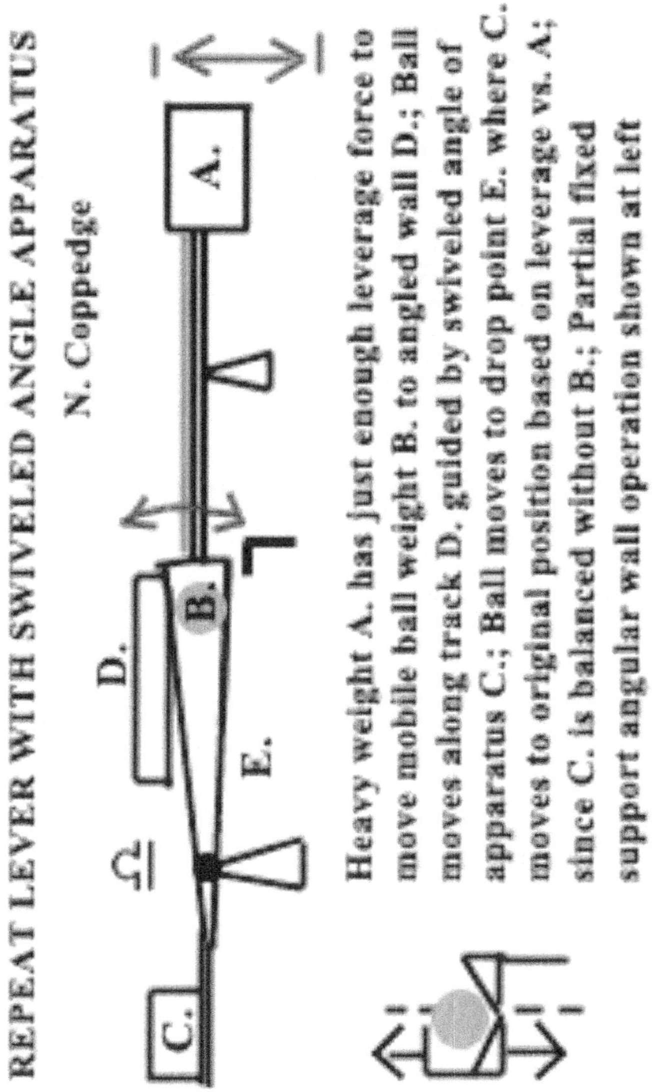

# PERPETUAL MOTION MACHINE DESIGNS & THEORY

## REPEAT LEVER WITH SWIVELED ANGLE APPARATUS
### N. Coppedge

Heavy weight A. has just enough leverage force to move mobile ball weight B. to angled wall D.; Ball moves along track D. guided by swiveled angle of apparatus C.; Ball moves to drop point E. where C. moves to original position based on leverage vs. A; since C. is balanced without B.; Partial fixed support angular wall operation shown at left

**COMPLICATED VARIATION OF THE REPEAT LEVERAGE APPARATUS**

*Nathan Coppedge*

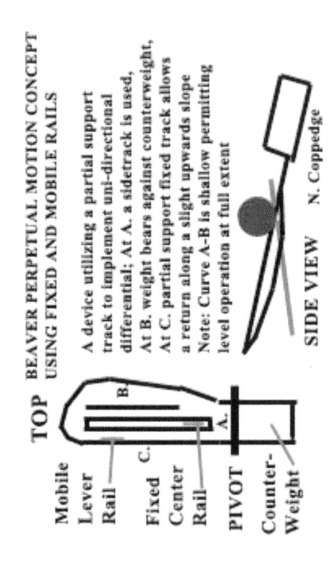

**BEAVER TYPE / FIXED RAIL SUPPORT VARIATION OF THE TROUGH LEVER**

# PERPETUAL MOTION MACHINE DESIGNS & THEORY

## TROUGH DEVICE USING SPIRAL-CURVED SLOPE

A device modifying an earlier principle for gain on differential support using a counterweight, in which as earlier a single ball weight is intended to operate, first upon a full trough, lacking support and therefore taking full weight of the mobile ball weight, and secondly upon a mobile half-trough partially supported by a fixed secondary half-trough

Nathan L. Coppedge

SIDE VIEW

counterweight

During the cycle, the ball weight moves from position A1 and A2, where the trough takes the weight, to positions B1 and B2 where the half-trough provides support

TOP VIEW

1.5 troughs, curved and mobile

fixed half-trough

arrows show cycle

**TROUGH LEVER VARIATION USING A CURVED LEVER / "BEAVER DEVICE"**

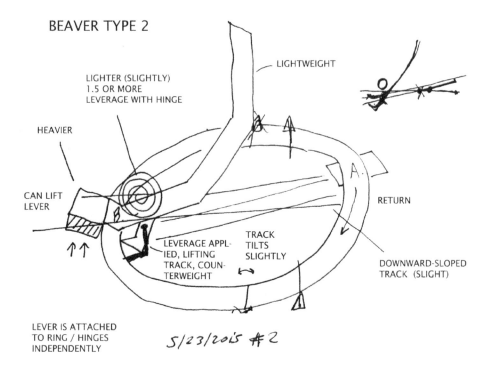

WORKING METHOD:
1. COUNTERWEIGHT IS HEAVIER (SLIGHTLY).
2. INITIAL LEVERAGE RATIO EXCEEDS RETURN RATIO.
3. B. BEGINS LOWER THAN A.
(VERY SLIGHTLY)---DIFFERENCE ACCOMODATED BY DISTANCE A. MOVES VIA LEVERAGE LATER

CAN GAIN HEIGHT A-B BECAUSE B BEGINS LOWER INITIALLY. NO GAIN IN HEIGHT B - A. LEVERAGE WORKS ON RETURN DUE TO SUPPORT RELATIVE TO APPLICATION FROM COUNTERWEIGHT.

# BEAVER TYPE 2

**PERPETUAL MOTION MACHINE DESIGNS & THEORY**

VARIATION ON THE REPEAT LEVER CONCEPT

ALTERNATION BETWEEN PARTIAL SUPPORT AND DOWNWARDS SLOPE

N. COPPEDGE

**SECANT TYPE REPEATING LEVER APPARATUS**

## DIFFERENCE PENDULUM DEVICE USING SPIRAL SLOPE AND STEEP OFFSET DROP POINT AND RETURN DRAG

At drop point (A) mobile ball weight has advantage against counterweight (B); However, according to design when ball weight reaches point (C) it is supported by a spiral track, decreasing effective weight, and causing counterweight (B) to act by dragging the weight upwards; By outwards angularity of the track and upwards pull, the ball weight is thrown in a spiral, until slope allows its full weight to apply at (C)::

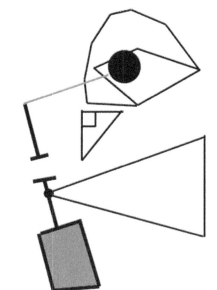

**REPEATING LEVERAGE VARIATION USING A "DIFFERENCE SPIRAL"**

## REPEAT LEVER EXPERIMENT

It is clear from this experiment that motion is not the sole factor of continuous momentum; For rolling ball at point (A) CAN return to point (C) at greater altitude, if it is allowed that the activated lever returns to a greater altitude; Again, the effectiveness of continuous momentum depends on specifics

(Approx. 2/3rds lever)

N. Coppedge

**REPEATING LEVERAGE EXPERIMENT**

*Nathan Coppedge*

## REPEATING LEVERAGE PERPETUAL MOTION MACHINE USING A HINGED STRUCTURE AND CLEVER USE OF RAMPS

Repeating leverage device making use of special ramp applications and a counterweight versus leverage application.

At point (A) rolling ball applies high leverage, operating the hinged lever structure and moving along section (B) which is sloped downwards. At the point beyond the hinge, the counterweight begins to exert greater force, and the ball weight is thrown upwards, at (C). A side ramp at (C) permits the ball weight to return to point (A) where the entire structure of the apparatus has returned to the initial position.

A counter-weight with approximately 3X mass at 1/2 hinge distance allows this machine to operate, when the structure is lightweight. In other cases both the counterweight and ball weight must be heavy to accomodate a heavy structure

*Nathan L. Coppedge*

**REPEATING LEVERAGE VARIATION USING A HINGE TO INCREASE SUPPORT-TO-DISTANCE**

**PERPETUAL MOTION MACHINE DESIGNS & THEORY**

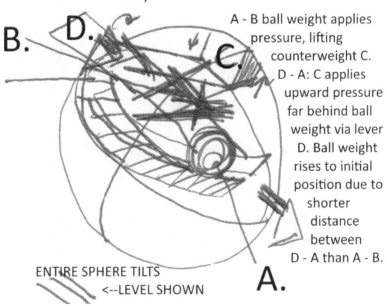

TILTING ORB / BOBBING SPHERE 1

A - B ball weight applies pressure, lifting counterweight C.
D - A: C applies upward pressure far behind ball weight via lever D. Ball weight rises to initial position due to shorter distance between D - A than A - B.

ENTIRE SPHERE TILTS
<--LEVEL SHOWN

**REPEATING LEVER SPHERE MAKING USE OF FIXED SUPPORT, COUNTER-WEIGHT, AND UPWARDS PRESSURE UPON EQUILIBRIZED LEVER**

Nathan Coppedge

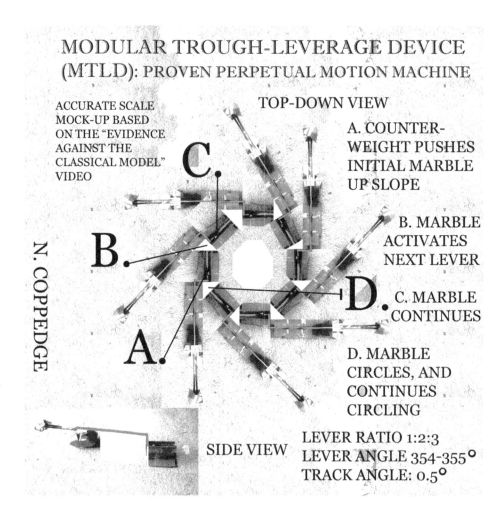

## MODULAR TROUGH LEVERAGE

### MOSTLY PROVEN BY EXPERIMENT

# PERPETUAL MOTION MACHINE DESIGNS & THEORY

## Tilt-Motor Perpetual Motion Concept

Original concept for a rotary device in which a weighted cone rolls around a swivel, activating successive pressure plates or "keys" operating levers. The levers in turn apply upwards pressure at a 90 degree angle on the track behind the cone. Since the track swivels downwards towards the portion weighted with the rolling cone, the upwards pressure is designed to create a continuous slope which follows the cone as it activates successive pressure plates.

Because the pressure plates are located outside the perimeter of the track, the cone's weight on the "wickets" on the active end of the levers only causes the pressure plate keys to be raised, rather than inhibiting movement by causing conflictive movement of the track. Metal "steps" are attached to the pressure plate keys in order to assure that the cone is activating one pressure plate at a given moment, which is meant to be sufficient to allow continuous motion.

Nathan L. Coppedge

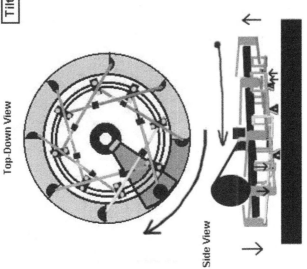

Top-Down View

Side View

## TILT MOTOR
### HIGH POWER, LOW PROOF

# THE "NOT-IF-BUT-WHEN MACHINE"

This device consists of four identical interconnected units, and a metal ball weight moving upon them.

The moving tracks marked with the hatches are lever-aged by counterbalance across the hatches from the shallow end.

The effect, permitted by the steep angle of the far end and fixed horizontal support on the rising, shallow end, is to propel the ball:

(1) activating the bar lever, with a very slight remaining downwards slope...
(2) up slightly, for a longer distance as necessary...
(3) continuing the process four times...
(4) Completing the loop and permitting perpetual motion (if built properly).

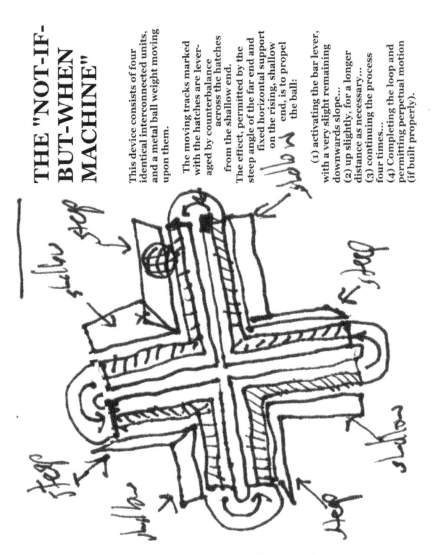

**EXCLLENT CONCEPT, MOSTLY PROVEN (REMEMBER THE COUNTERWEIGHTED LEVERS!)**

## THE "NOT-IF-BUT-WHEN MACHINE" 2

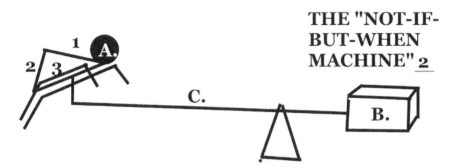

Ball weight A. follows course 1-2-3 by action of approximately equal counterweight B. through lever C. Lever C. has proven proportions of about 2.2 - 2.8 / 1, where 1 represents the length of the short end. Motion begins with an upwards angle 1, lowered by full application, slight downwards slope 2, and heavier downwards slope 3 permitting angle to act through B, to C, upon A.

## SOME VARIATION OF THIS DESIGN PROBABLY WORKS, BUT FINDING IT MAY BE TRICKY

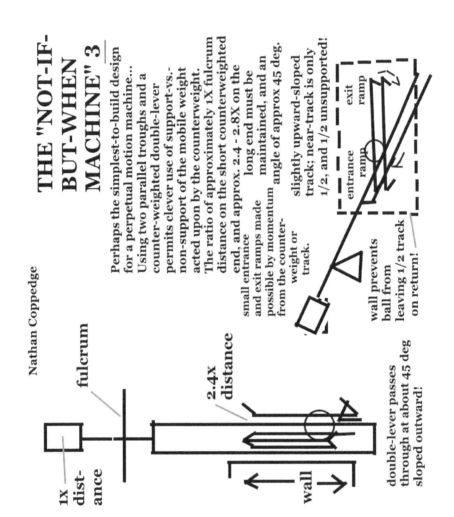

# THE "NOT-IF-BUT-WHEN MACHINE" 3

Perhaps the simplest-to-build design for a perpetual motion machine... Using two parallel troughs and a counter-weighted double-lever permits clever use of support-vs.-non-support of the mobile weight acted upon by the counterweight. The ratio of approximately 1X fulcrum distance on the short counterweighted end, and approx. 2.4 - 2.8X on the long end must be maintained, and an angle of approx 45 deg. small entrance and exit ramps made possible by momentum from the counter-weight or track.

slightly upward-sloped track; near-track is only 1/2, and 1/2 unsupported!

entrance ramp

exit ramp

wall prevents ball from leaving 1/2 track on return!

fulcrum

1X distance

2.4x distance

wall

double-lever passes through at about 45 deg sloped outward!

**THIS IS AN EXCELLENT DESIGN WITH A MOSTLY PROVEN PRINCIPLE, MUCH SIMPLER TO BUILD THAN MOST**

## "NOT-IF-BUT-WHEN" MACHINE #4

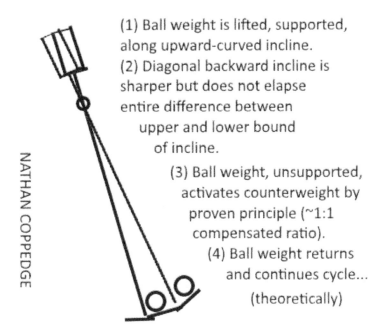

(1) Ball weight is lifted, supported, along upward-curved incline.
(2) Diagonal backward incline is sharper but does not elapse entire difference between upper and lower bound of incline.
(3) Ball weight, unsupported, activates counterweight by proven principle (~1:1 compensated ratio).
(4) Ball weight returns and continues cycle...
(theoretically)

**THIS IS A MOSTLY PROVEN DESIGN AND THE FIRST TO PROVE THE RETURN-AT-EQUAL-ALTITUDE PRINCIPLE**

## NOT-IF-BUT-WHEN MACHINE #5

WEIGHT RATIO 1:1 COMPENSATED.
LEVER RATIO: APPROX.
1 : 2 - 3
OR
1 : 2.4 - 3.2

BALL WEIGHT ENTERS AT PT. A WHERE STEEP FIXED SUPPORT ACTS AGAINST MODERATE UPWARD LEVERAGE FROM COUNTERWEIGHT. AT PT. B DIRECTION CHANGES AND LEVERAGE IS MORE STEEP FROM COUNTERWEIGHT, WHILE FIXED SUPPORT IS SHALLOW, REDUCING RESISTANCE TO UPWARD MOTION FROM COUNTERWEIGHT. WHEN BALL WEIGHT REACHES PT. C, MOMENTUM TRANSFERS BALL WEIGHT TO NEXT OF THREE ADDITIONAL UNITS, REPEATED HORIZONTALLY AS ALREADY SHOWN, SO CYCLE CONTINUES... BACK TO BEGINNING

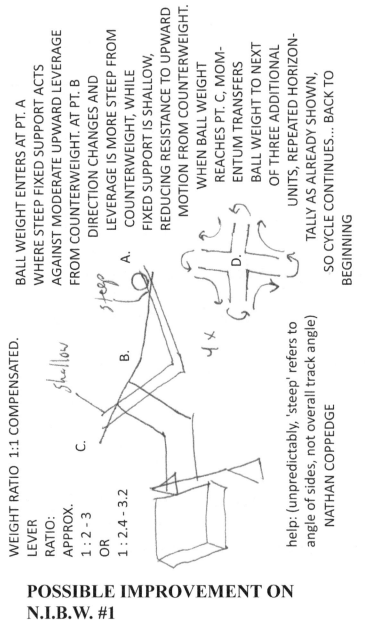

help: (unpredictably, 'steep' refers to angle of sides, not overall track angle)

NATHAN COPPEDGE

**POSSIBLE IMPROVEMENT ON N.I.B.W. #1**

## PERPETUAL MOTION MACHINE DESIGNS & THEORY

## NOT-IF-BUT-WHEN MACHINE #6

NATHAN COPPEDGE

A. WEIGHT RATIO 1:1 COMPENSATED LEVER RATIO 1: 2.4 - 2.7+

B.

C.

AT PT. A COUNTER-WEIGHTED CIRCULAR WIRE (LIGHTER COLOR) APPLIES ANGLED PRESSURE AGAINST BALL WEIGHT. BALL WEIGHT RISES BECAUSE IT IS SUPPORTED.

PROCESS CONTINUES UNTIL BALL REACHES APEX B, AT WHICH POINT CIRCULAR WIRE ANGLE CHANGES TO A STEEP UPWARD INCLINE. HOWEVER, BALL APPLIES SIGNIFICANT PRESSURE ON WIRE BECAUSE IT IS NOW ALMOST FULLY UNSUPPORTED DUE TO STEEP ANGLE C. SINCE BALL AND COUNTERWEIGHT ARE APPROXIMATELY EQUAL IN COMPENSATED MASS BALL WEIGHT SINCE IT HAS GREATER LEVERAGE CAN CONTINUE PROCESS.

## LONG-DELAYED DESIGN
## ADAPTED TO CIRCULAR WIRE

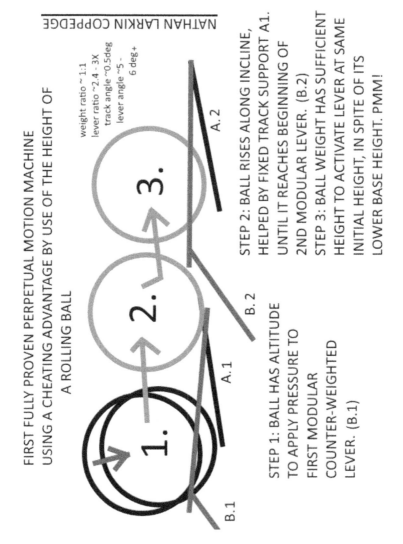

**FIRST FULLY-PROVEN PERPETUAL MOTION MACHINE.**
RATIOS ARE ALL ABOUT THE SAME AS IN THE MODULAR TROUGH LEVERAGE DEVICE.

**PERPETUAL MOTION MACHINE DESIGNS & THEORY**

Nathan Coppedge

# THE ESCHER MACHINE

**C.** Master Angle 2: marble rolls upwards again, using a differently-directed master angle

**D.**
Ramp 2: Using altitude from Master Angle 2, marble returns to Master Angle 1

**B.**
Ramp 1: A downwards motion is possible due to the gain in height

**A.** Master Angle 1: marble rolls upwards using a horizontal slope

NATHAN COPPEDGE

**SIMPLE, SUBTLE OPTION**

PERPETUAL MOTION MACHINE DESIGNS & THEORY

## THE ESCHER MACHINE

WHEN SLOPE (H) > SLOPE (V), DIRECTION OF MOTION IS SOMETIMES VERTICAL WHEN GRAVITY IS DIRECTED HORIZONTALLY

Direction of Motion

G Vector     G-F     Force Vector
             Differ-
             ential
                     NATHAN
                     COPPEDGE

## ESCHER MACHINE PHYSICS

**85**

*Nathan Coppedge*

## COQUETTE

**A CONCEPT I DREAMED UP IN 2007**

**PERPETUAL MOTION MACHINE DESIGNS & THEORY**

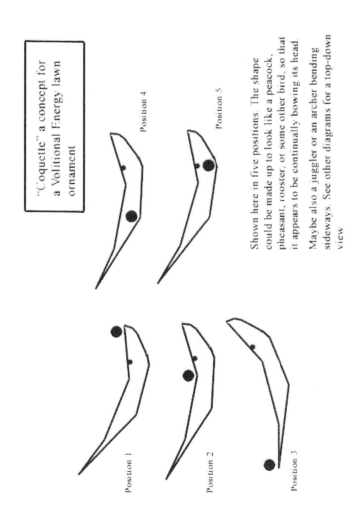

**OPERATION OF THE COQUETTE (THEORETICAL)**

## COQUETTE: A Volitional Energy Lawn Ornament Concept
Eucaleh Terrapin, Inventor

In this device, which may function as a lawn ornament, or if suitable in that role, also as a mechanical device, a structure of track or tubing is built over a fulcrum, like a see-saw divided into three parts that meet at each end.

One end is considerably shorter than the other, that is so that Point A is much closer to the fulcrum than Point B. A single ball-bearing type weight is used, so that when it is placed at Point A the natural slope leads downwards some ways towards Point B. When the ball bearing reaches a point between A and B where its weight would account by leverage for the shorter end of the pivoting structure, the angle is such that if it did not the slope would be prohibitively steep, but instead the weight of the ball-bearing extends the slope until it reaches point B

At Point B, the ball bearing may take one of two routes, which are identical. It may follow 1 or 4 downwards some length until Point B begins to rise again. Then at 2 or 5 the slope is such that it would be prohibitive if Point B. were at its lowest. Finally at Seg. 3 or 6 the slope is down into Pt. A

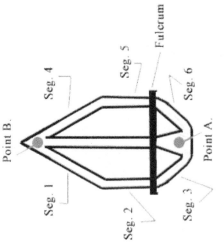

A Top-Down View

Point B.

Seg. 1
Seg. 2
Seg. 3
Seg. 4
Seg. 5
Seg. 6

Fulcrum

Point A.

Note: because the slope averages an upwards grade between Point A. and Point B. if the device did not pivot, by the time the ball bearing returns to Point A. it may travel downwards

**PRIMARY OPERATION OF THE COQUETTE**

## PERPETUAL MOTION MACHINE DESIGNS & THEORY

### Coquette Type 2

Tilting device relying upon a differentiation in leveraged weight application, vis. at Point A the heavy weight is resisted since, although it has greater bearing, it applies less leverage than the counterweight, however at point B the light weight is resisted via strong counterposed leverage combined with a heavy weight at less leverage distance than the rolling ball. By following a circular or ovular track the ball weight thus finds points within which a forward impetus is reinforced by the tendency of the device to tilt where the ball bears down, while returning with force to an anteceding position, the principle of light leverage (counterpointed by leverage application) counterpoints heavy weight application at a point of minimal leverage.

A.

light counterweight w/ leverage

heavy weight with low leverage / torque

circular / ovular track

b.

this design is based on bisected metal supports for atmospheric lights

TECHNIQUES

LEFT: key slope points corresponding to major lever applications

|| Eucaleh Terrapin ||

**COQUETTE TYPE 2 (NOTE DISK)**

## PERPETUAL MAGNET CONCEPTS

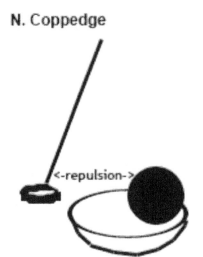

**MAGNET DESIGN 1**

A circular magnet is attached at its midpoint to a tether, above a spherical bowl in which a larger metal ball is made to roll; When the magnet approaches the ball, the ball rolls, applying its weight to rotation

**MAGNET MACHINE TYPE 1: HORIZONTAL APPLICATION OF WEIGHT AGAINST AN UPWARDS CURVE IS DESIGNED TO CREATE A PERPETUALLY-ROTATING PENDULUM. NOTE THE HORIZONTAL CURVE OF THE BALL IN THE MIDDLE**

[SEE ALSO GRAVITY MOTORS, LATER]

N. Coppedge

**PERPETUAL MOTION MAGNET 1**

Modification

If a single magnet is both attractive and repulsive, two magnets can be mounted together oppositely in a cylindrical form, with the mid-point being aligned with the mid-point of the larger ball weight, creating the same effect of repulsion

<-repulsion->

**MAGNET MACHINE TYPE 1, VARIATION 1: COMPRESSING MAGNETS TOGETHER MAY GENERATE MORE OR EXCESS FORCE, BUT RUNS THE RISK OF DEGRADING THE MAGNETS**

## BEZEL WEIGHT DEVICE

TRACKED APPARATUS USING A BEZEL-WEIGHT AND ROLLING BALL

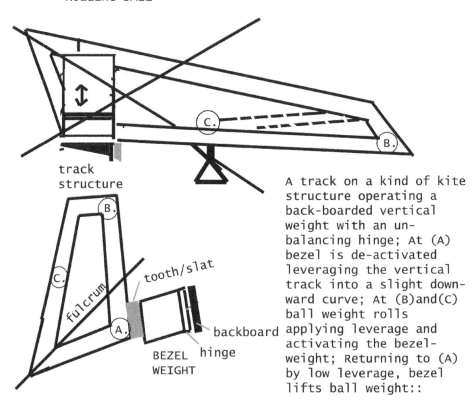

track structure

A track on a kind of kite structure operating a back-boarded vertical weight with an un-balancing hinge; At (A) bezel is de-activated leveraging the vertical track into a slight downward curve; At (B) and (C) ball weight rolls applying leverage and activating the bezel-weight; Returning to (A) by low leverage, bezel lifts ball weight::

### BEZEL WEIGHT TYPE 1/1

BEZEL DEVICES MAKE USE OF A WEIGHT AGAINST A BRACKET TO CREATE SPRING-WORK INDEFINITELY

# PERPETUAL MOTION MACHINE DESIGNS & THEORY

## GRAVITY MOTOR

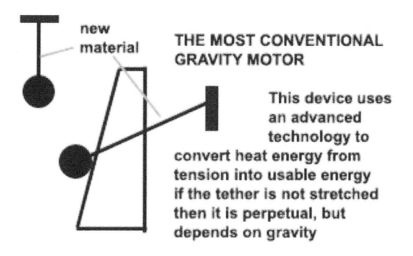

EVEN BASIC GRAVITY MOTORS MAY REQUIRE SOME SORT OF ADVANCED TECHNOLOGY, SUCH AS NANO-ROPES, TO WORK

## PMM CONCEPT "GRAVITY CLOCK" INCORPORATING A HORIZONTAL DISCUS

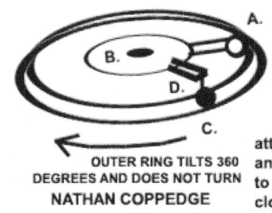

Ball bearing (A) creates slope which transfers through central lightweight disk (B) to mobile ball-weight (C) attached by member and tether (D), aimed to create in this case clockwise motion

OUTER RING TILTS 360 DEGREES AND DOES NOT TURN

NATHAN COPPEDGE

## PMM CONCEPT "GRAVITY CLOCK" INCORPORATING A HORIZONTAL DISCUS : VARIATION ONE

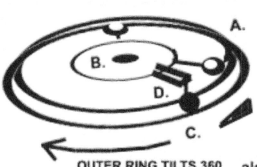

A design similar the earlier form of gravity clock except involving a 'roller-skate' double-ball-bearing implementation also using a single rolling ball weight for motion

OUTER RING TILTS 360 DEGREES AND DOES NOT TURN

NATHAN COPPEDGE

**GRAVITY-CLOCK GRAVITY MOTOR CONCEPT. SIMILAR DESIGNS HAVE FAILED TO PROVIDE ENOUGH ENERGY FOR MOTION, E.G. TILT MOTOR EXPERIMENT 2 AND 3.**

# SPIRAL WHEEL

## VERTICAL WHEEL USING SPIRALS AND DOUBLE-DIFFERENCE

A vertical wheel made of spirals and joined from the exterior (not pictured), using two fixed horizontally rotating pendulums; The first [A] is heavier, acting on the second [B], which has enough weight to resist the wheel, providing upwards force against the first pendulum; The rotary motion of the first pendulum is meant to counteract the second, creating a circular motion of the wheel

[SIDE VIEW]

[FRONT VIEW]

Nathan Coppedge

**VERTICAL WHEEL USING PENDULUMS IN AN EFFORT TO CREATE COUNTERVAILING SPIRALS**

## SECOND ATTEMPT AT A CONVENTIONAL WHEEL

HERE AN OUTWARDS-PROTRUDING SUPPORTING TRACK HAS BEEN CLEVERLY ATTACHED TO WHAT IS IMPORTANTLY A FIXED AXLE. THE WHEEL IS PERMITTED TO ROTATE BY A CIRCULAR ENGAGEMENT WITH THE FIXED AXLE. ADDITIONAL MOMENTUM IS SUPPOSED TO BE PROVIDED BY THE OUTWARDS-SLOPED BALL WEIGHT (AT LEAST ONE AT A TIME), AND, ADDITIONALLY, A VERTICAL BAR PUSHES THE OPPOSITE WEIGHTS INWARD TO REDUCE RESISTANCE.

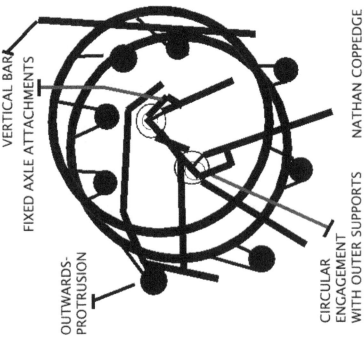

**VERTICAL WHEEL 2
(CONVENTIONAL WHEEL)**

# PERPETUAL MOTION MACHINE DESIGNS & THEORY

CONVENTIONAL WHEELS / MY RENDITION OF THE FAMOUS BESSLER WHEEL

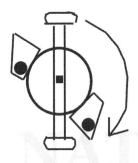

1. The rightmost ball weight begins to fall, by superior leverage.

2. The bar weight begins to equilibrize

3. As the bar weight equilibrizes, the ball weights begin to roll to the right.

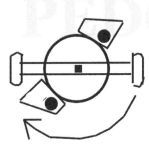
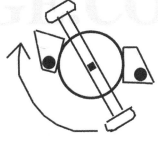

**NATHAN COPPEDGE**

4. The bar weight is equilibrized, but the ball weights are unbalanced.

5. The ball weight begins to fall again, returning the device to position 1.

## FIRST RENDITION OF THE BESSLER WHEEL CONCEPT, BASED ON SUR-MISES

## PENDULUM DEVICES

**DIFFERENCE PENDULUM DEVICE USING SPIRAL SLOPE AND STEEP OFFSET DROP POINT AND RETURN DRAG**

At drop point (A) mobile ball weight has advantage against counterweight (B); However, according to design when ball weight reaches point (C) it is supported by a spiral track, decreasing effective weight, and causing counterweight (B) to act by dragging the weight upwards; By outwards angularity of the track and upwards pull, the ball weight is thrown in a spiral, until slope allows its full weight to apply at (C)::

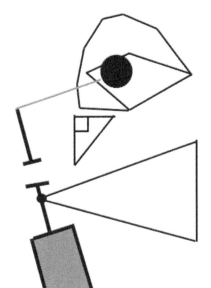

**PENDULUM TYPE 1**

# PERPETUAL MOTION MACHINE DESIGNS & THEORY

## HORIZONTAL SPIRAL DEVICE
### USING A DUAL-WEIGHTED CORD

Position A. Outer ball weight follows the inward curve of the spiral, which is steeper than the upward lateral curve supporting the weight; Position B. Outer ball weight continues to follow the curve, as the central fulcrum swivels the counter-weight; Position C. The outer ball weight follows an incline which is more vertical than the sum of the lateral and spiral, so that the weight returns to the start point::

## PENDULUM TYPE 2

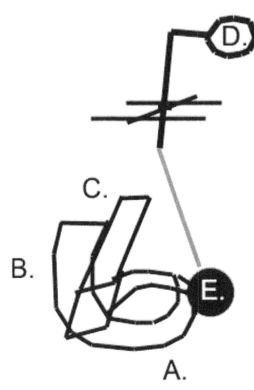

**Counterweighted Pendulum, Type 3**

At (A) downwards slope pushes ball weight E. outwards, At (B) slope becomes gentle, allowing an upwards climb. At (C) ball drops, lifting counterweight (D.) and pushing ball weight (E.) outwards. Proportionality seems possible because ball weight (E.) is heavier than counterweight (D.)

horizontal support allows (D.) to lift (E.) ⬅

**PENDULUM TYPE 3**

PERPETUAL MOTION MACHINE DESIGNS & THEORY

## SPINNING TOP DEVICES

**PERPETUAL MOTION "SPINNING TOP"-TYPE CONCEPT #1**

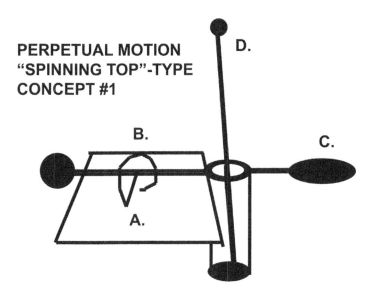

At point A. inward-curving spiral member causes main shaft to rise upon a slope. Main shaft loses altitude but spiral gains altitude. It is eased by counterweight C. The ramp may tilt, allowing the main shaft to swing back to position A. In general, the motion allows pin D. to spin, generating energy.

SPINNING TOP #1

## PERPETUAL SPINNING TOP TYPE #2

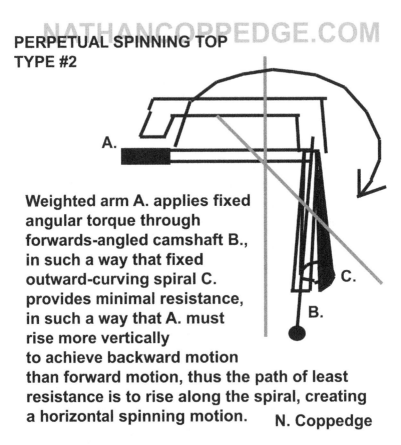

Weighted arm A. applies fixed angular torque through forwards-angled camshaft B., in such a way that fixed outward-curving spiral C. provides minimal resistance, in such a way that A. must rise more vertically to achieve backward motion than forward motion, thus the path of least resistance is to rise along the spiral, creating a horizontal spinning motion.   N. Coppedge

**SPINNING TOP #2**

PERPETUAL MOTION MACHINE DESIGNS & THEORY

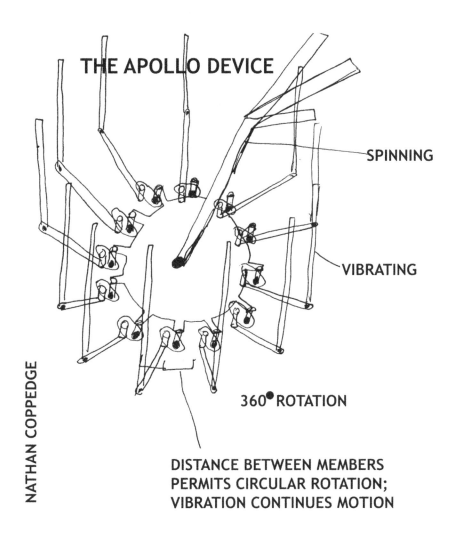

APOLLO DEVICE -- Perhaps existed in Ancient Greece.

*Nathan Coppedge*

# PERPETUAL MOTION MACHINE DESIGNS & THEORY

## ODDITY #1

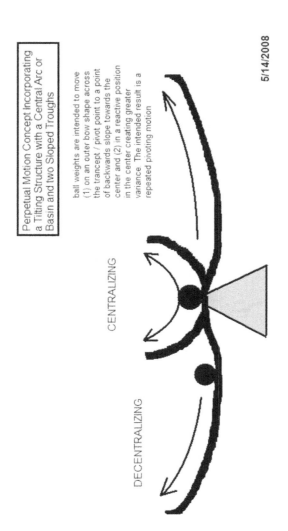

**PERPETUAL MOTION BALANCE [?]**

## ODDITY # 2:

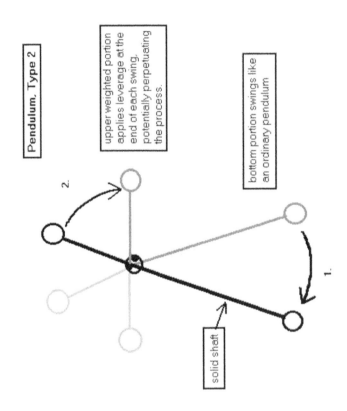

**PENDULUM TYPE 2 [?]:
UPPER PIECE MOVES L/R SLIGHTLY**

**PERPETUAL MOTION MACHINE DESIGNS & THEORY**

ODDITY #3

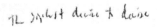

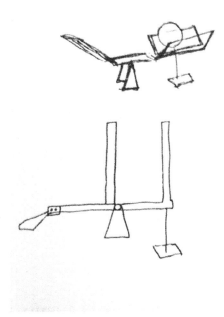

**DEVICES FOR GENERATING
ENERGY IN PUBLIC PLACES**

## ODDITY #5

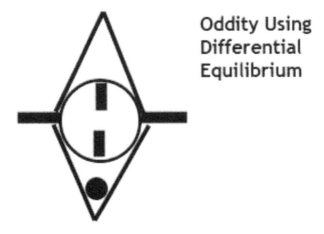

Oddity Using Differential Equilibrium

Theoretically in a divice using the same parts as above, momentum contributes to four "shift" promoting continuous motion, according to a principle of equilibrious motion---

[FOR #4 SEE SIDEWAYS LEVERAGE UNDER REPEATING LEVERAGE]

**[SIC] DEVICE.**

**A DEVICE MAKING USE OF EQUILIBRIUM (PERHAPS SEVERAL DESIGNS)**

PERPETUAL MOTION MACHINE DESIGNS & THEORY

## ODDITY #6

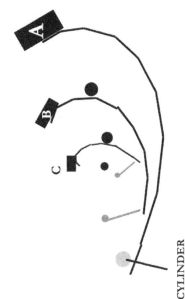

Cylinder enters basin A with no energy input; Rolls to tip of cylinder B at which point A begins to counteract with counterweight force; Cylinder rolls to tip of cylinder C, activating basin B counterweight, providing lift; C begins to operate, providing additional lift;

Nathan L. Coppedge

PARTIAL ENERGY CONCEPT [AUGMENTED ENERGY CONCEPT] USING TWO-DIMENSIONAL BASINS IN RECESSED CONFIGURATION

**DEVICE FOR ROLLING UPWARDS**

*Nathan Coppedge*

## "APERRATURES":

## APPLICATION OF PERPETUAL MOTION TO BUILDINGS CONSTRUCTION

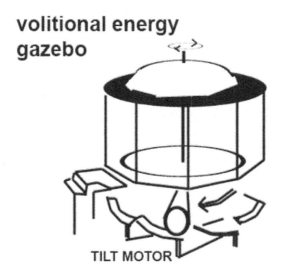

Perhaps the simplest type of aperrature (mobile building)

PERPETUAL MOTION GAZEBO

**PERPETUAL MOTION MACHINE DESIGNS & THEORY**

### APPERATURE RAMP (AUTOMATED RAMP) USING REPEATED VESCENSION

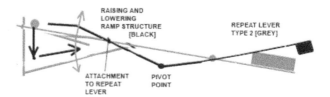

Implementation that would ostensibly allow automated ascension, via a principle of lever advantage versus counterweighted mass and momentum to overcome equilibrium, essentially a repeating lever for buildings; Use is made of a counterbalance, as with elevators

### PEPRETUAL MOTION RAMP

Nathan Coppedge

Aperrature (Mobile Building) Using Counter-Weighted, Ratcheted Tilt Motor

Here a free-floating Tilt Motor apparatus, augmented by counter-weight, operates a ratchet, creating a circular movement

**PERPETUAL MOTION RATCHET**

# HOW TO BUILD AN OVER-UNITY DEVICE

*Nathan Coppedge*

# PERPETUAL MOTION MACHINE DESIGNS & THEORY

"Trough Leverage Partial Perpetual Motion Machine"

"SMOT Without Magnets"

Parts:
1 X 1/8 in. diameter aluminum rod (approx. 12 in. long).
1 X long k'nex member (orange, black, or tan) approx. 6 in. X 1/8 in.
2 X medium k'nex members (red or yellow) approx. 4 in. X 1/8 in.
7 X medium-small k'nex members (blue or yellow) approx. 2 in. X 1/8 in.
1 X Additional optional k'nex member, easily replaced with wire
7 X half-circle k'nex connector pieces (yellow) approx. 1.5 in. X 1 in X 1/4in.
Duct Tape
Indefinite number of Amazon book-box quality cardboard of medium thickness.
1 X Medium (slightly larger) marble
1 X Standard marble
5 X pennies (not always necessary)
3 X packets of sculpee clay or small heavy block of wood.

Steps:

Attach the 1/8 in. diameter aluminum rod to the long k'nex member, wrapping tightly with as little duct tape as is necessary to make it secure. Remove the duct tape if bunching occurs, as this will reduce the stiffness of the member. Favor lightweight over heavy connections, and making sure that the entire length is about 16 - 18 inches.

Take two half-circle k'nex connectors, and attach 2 medium k'nex members, attaching one to each, close to the flat side. Now place a member through the hole in the connectors, with the previously attached medium connectors facing the same direction, with the flat ends of the pierced connectors up. Now securely fasten the bunched medium members to the earlier-made long member, so that the hole in the connectors is located at exactly 1/4 the length of the entire long member (preferably on the side with the k'nex member---as opposed to the aluminum rod---so that there is a more stable connection) with the majority of the length opposite the members just attached. Adjust the tape if necessary to make sure that a member can freely rotate inside the connectors which are attached to the lever.

Now take two more half-circle connectors, and attach 4 medium-small members, 2 per connector, in each case forming a symmetrical 90-degree angle on the connector. Place a member of any adequate length through the secured lever-connectors and then support this new fulcrum joint with the attached medium-small members, by piercing the new half-circle connectors.

Attach two additional half-circle k'nex connectors to the ends of the taped unit (now a lever), positioning them so the flat sides are turned towards eachother, forming a narrow floating base. Then attach a medium (larger than standard) marble to this base by duct tape. Slight additional weight may be necessary later, but can be offset by re-positioning the lever forwards or back. Make sure the marble is secure, but that there are not excessive globs of duct tape, which would weigh down the machine. Later, during testing, it may be necessary to re-attach the marble using the same amount of duct tape, or to attach additional duct tape with small additional weights, or just duct tape serving as a weight.

Create a cardboard base of indefinite dimensions, at least 4 in. wide by 12 in. long.

Take more pieces of cardboard, cutting the cardboard to create two very slightly sloped panels, from 6 - 12 inches long (about 1/8 in. thick), and towards the outer end relative to the position of the lever, incorporating a slight slope of only 0.5 degrees which extends for about 4.5 - 5 inches horizontally and has a height of about 2-3

## PERPETUAL MOTION MACHINE DESIGNS & THEORY

inches, such that the lever passes through at an angle of 6-5 degrees below level (this can be adjusted by raising or lowering the fulcrum or hinge), the steeper angle being the angle at which the upper surface of the lever meets the beginning of the 4.5 - 5 inch long slot in the track.

Alternately, cut sheer horizontal pieces of the dimensions 8 inches by 3 inches, and later cut the cardboard to approx. 0.5 degrees upwards slope after it is taped in place.

Tape the cardboard track members to the cardboard base, allowing room for the lever to pass through freely. Make sure the track members are taped securely by placing tape laterally over the vertical duct tape pieces, once the track is in position for the lever. Make sure that the gap is narrow enough to support a marble (when it is stabilized) along the 0.5 degree upward slope, and wide enough to accommodate the lever passing through the slot.

Test the position of the lever, allowing a gap thick enough to accommodate the long end of the lever.

Three of the medium k'nex members can be used to stabilize the separation between the two sides of the cardboard track. Tape the three members closely together, so they form a 3 X 1 unit on their shortest length. Cut two approx. 2 in. X 4 in. cardboard members, lower the lever into the slot, These should be fastened with duct tape to the bottom of the outer sides of the portion of the track unit which does not include the slope, without obstructing the lever. Then tape the cardboard members to the two side lengths of the 3 X 1 member, inside and outside of the slotted arrangement, but without obstructing the lever. The 3 X 1 serves as a separator and stabilizer between the two sides of the track, keeping the gap somewhat less than the width of a standard marble.

Tape cruder outer horizontal cardboard panels around the active track portion of the unit, e.g. on the outside side of the actual track portion where the smaller marble will be rolling, allowing some visibility, but allowing for greater control of the marble when it is being placed, and to prevent the marble from flying out when the lever is in the wrong position. Now the track portion is complete, assuming that the altitude of the lever can be adjusted, and that the angularity of the track is very slight (only about 0.5 degrees, and extending for about 4.5 - 5 inches).

Tape the k'nex fulcrum supports with the lever attached between them from an earlier step to some sort of solid base, about three inches off the surface being used, and ideally with at least 2 sq. inches of surface area at 3 in. height for attaching with tape. Packets of sculpee clay are ideal, but other things such as blocks of wood can be substituted. It is best if this item is solidly attached, because otherwise it is possible to inadvertently lose appropriate angularity, or even to damage other elements of the device. Tape the lever to this second base when it is ascertained that the angularity of the lever at 2 - 3 X the distance of the center of the counterweight is equal to 6 - 5 degrees below level, with 6 degrees occurring at the beginning of the (0.5 degree) and 4.5 - 5 inch long upward-sloping track.

Alternately, test the position of the lever with attached counterweight by trial and error, locating the angle at which the smaller marble moves when placed at the beginning of the finished track.

When the unit is complete, additional experimentation may be required to find a repeatable process, but when all the steps are followed, the process should be significantly repeatable. Here is a troubleshooting guide for the final success stage:

(1) If the lever won't move the marble upwards, maneuver the lever up and down, to see if there is any obstruction in the slotted track. There should be zero obstruction, and the marble should still have support from both sides of the track. Be careful to avoid bumps in the angularity of the upwards slope of the cardboard, because this changes the effective angle of operation. Adjust the

# PERPETUAL MOTION MACHINE DESIGNS & THEORY

construction or lean and rotation of the track and track base if necessary.

(2) If there is no obstruction and the counterweight will still not lift the marble at 2 - 3 X counterweight distance on the opposite end of the fulcrum, or if the counterweight catapults the marble vertically instead of pushing it horizontally, then it is time to add or subtract weight from the counterweight. Did you use a large marble, instead of a medium one, for the counterweight? If so, that is too heavy to use. Try carefully adjusting the weight of the counterweight with pennies, without the large marble. If you used a medium marble, but the smaller marble won't move, and there is absolutely no obstruction in the track portion, then try adding one or two pennies. It is possible that you used heavier plastic members, or attached the aluminum rod differently. You can also try once again adjusting the distance between the fulcrum and the slotted track.

(3) Now, experiment by trial and error. In earlier stages, it should have been certain that the counterweighted lever could push the marble horizontally. Now, with a very slight upward angle, the same process should repeat, and it should become clear that the lever returns back to its height after being operated. It is also clear that every part which operates can return to its beginning altitude, that is, when stops are placed on the range of the lever, and when the mobile marble is stopped when it drops to its beginning altitude. Over-unity!

A perpetual motion machine requires crooked levers (that is, levers designed with a double-bend somewhere between the fulcrum and the 2 X distance to allow the tracks to cohere), and a duplication of the above concept eight times, maintaining identical altitudes, with every lever end chopped at or immediately after the end of the upwards track, and with the base of every upwards track positioned immediately beneath a previous track, with an angular difference in orientation of perhaps 45 degrees for every unit that follows, creating an octogon, or perhaps with a use of short-range assistive slopes, 180 degrees and only 2 X modularity.

*Nathan Coppedge*

# ADDITIONAL THEORY

*Nathan Coppedge*

**PERPETUAL MOTION MACHINE DESIGNS & THEORY**

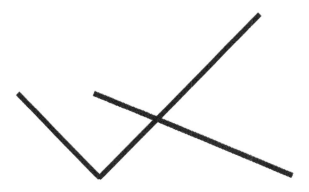

With 2/3rds lever ratio, supporting track should be at 1/2 the angle of the lengthier end, differently directed or even more shallow!

This allows for dual-directional lever advantage when the mobile weight is fully acting on the lever through half of its motion

[ADDITIONAL THEORY 1]

*Nathan Coppedge*

The lengthier the lever, the slighter its upward and downard movement should be!

The upward and downward movement of any mobile weight is thus correspondingly shallow.

**[ADDITIONAL THEORY 2]**

# PERPETUAL MOTION MACHINE DESIGNS & THEORY

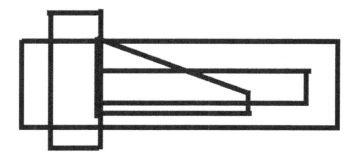

Working with 2-dimensional shapes will result in different optimal weight configurations for each lever, such as the above, shown from above!

(See if you can find the fulcrum points where they balance)

[ADDITIONAL THEORY 3]

*Nathan Coppedge*

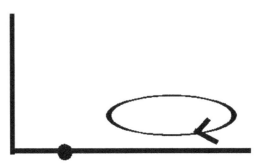

Maximizing a mobile weight's effect on the counterweight involves maximizing the area of effect!

This involves reducing the length of the counterweight relative to its mass.

**[ADDITIONAL THEORY 4]**

# PERPETUAL MOTION MACHINE DESIGNS & THEORY

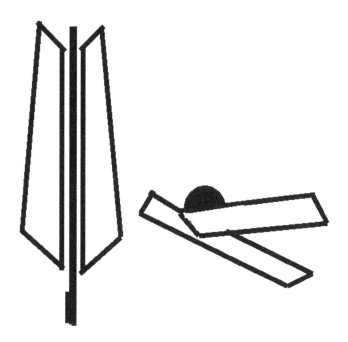

The method on the left is an improvement on the method on the right for lifting a ball along a supporting track.

About 75% vs. 50% support in nearly horizontal arrangements.

[ADDITIONAL THEORY 5]

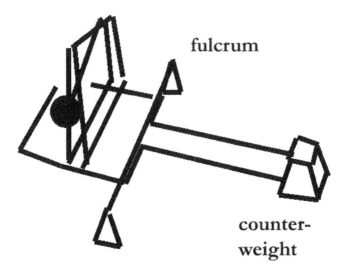

An interesting pendulum arrangement maximizing horizontal motion, provided that the upward and downward permitted motion is very slight. One can experiment with the angle of the counterweight lever and supporting board. Note that some of the ratios are maximized.

**[ADDITIONAL THEORY 6]**

A more advanced maximized pendulum. Note that the above arrangement may be weighted in ratio to length, and the motion other than the pendulm is again very slight. This arrangement has been recorded to make up to 418 swings.

[ADDITIONAL THEORY 7]

3-D design

In the case of the Escher Machine, three principles are combined: (1) Momentum from the backboard, (2) Horizontal displacement, and (3) Upward motion. Some evidence has been provided in experiments.

[ADDITIONAL THEORY 8]

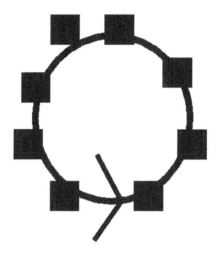

The theory of modular units says that if each modular unit maintains the same altitude and can cause motion from rest, then the cycle perpetuates much like the proverbial process of recycling.

[ADDITIONAL THEORY 9]

*Nathan Coppedge*

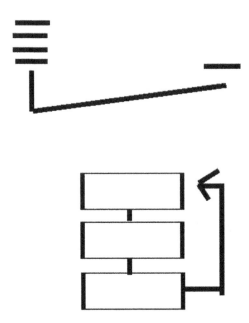

How to make use of modular units? There are at least two ways: (1) If one or many units can trigger all the other units at no expense, and (2) If each unit can trigger the next unit at no expense.

[ADDITIONAL THEORY 10]

## THEORIES OF EFFICIENCY

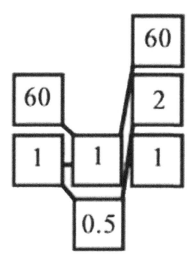

If a device has a principle of energy such as slope or a chain reaction, it still needs a principle of efficiency, such as a method of "cheating".

[ADDITIONAL THEORY 11]

*Nathan Coppedge*

Modular units are a deceptively difficult method to achieve high energy.

Most of the time such energy is factored out because the units are actually dual-directional.

**[ADDITIONAL THEORY 12]**

Efficient energy may be had through a mixture of providing momentum and using efficiency factors to "cheat" physics.

**[ADDITIONAL THEORY 13]**

The principle of equilibrium
is a case in point for efficient
energy!

Balancing scales will essentially
move from rest, raising the
prospect that physics can be
"cheated".

**[ADDITIONAL THEORY 14]**

# PERPETUAL MOTION MACHINE DESIGNS & THEORY

For example, if two equal weights are positioned on either end of a balancing scale, you will find that one weight will be lifted to a higher altitude if it is supported by a slope.

[ADDITIONAL THEORY 15]

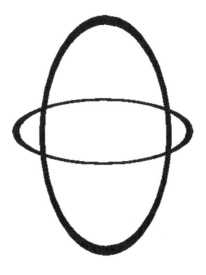

There are essentially two choices:
to create a horizontal loop,
or to create a vertical loop.
Although, in some cases such
as Motive Mass Iteration 2,
this pattern may be circumvented.


**[ADDITIONAL THEORY 16]**

# PERPETUAL MOTION MACHINE DESIGNS & THEORY

Big Wheel, Little Wheel: Although vertical wheels are tempting, in every case it is easier to create horizontal than vertical motion when contending with gravity.

[ADDITIONAL THEORY 17]

*Nathan Coppedge*

In order to overcome the problem of proportionality, there must be at least one efficiency,
such as:
(1) A horizontal circle,
(2) An imbalance, or
(3) A method of "cheating" such as a supporting track.

**[ADDITIONAL THEORY 18]**

Combining multiple methods of cheating is the usual method, whereas imbalance and horizontal circles are often used as reserve methods.

I say this tentatively, because in my view the science of perpetual motion is still rather young.

At least, functional perpetual motion.

[ADDITIONAL THEORY 19]

*Nathan Coppedge*

Methods of cheating include:
(1) Use of a balance.
(2) Use of a supporting track for half of all motion.
(3) Using lesser distance for heavy weights.
(4) Using greater distance for lighter weights.
(5) Vertical support such as a pendulum arrangement or fulcrum.
(6) A mixture of elements.
(7) Combined principles.

Notice, no where here do I mention friction!

**[ADDITIONAL THEORY 20]**

## THE FOUR PROOFS OF PERPETUAL MOTION

(1) Friction does not eliminate motion where motion is permitted.
(2) Objects can chain-react in a circle, as shown by dominoes; wheels can turn.
(3) Objects can chain-react using higher and higher altitudes. Energy can be created.
(4) Dominoes can accelerate, so friction does not stop everything!

[ADDITIONAL THEORY 21]

*Nathan Coppedge*

## FOUR BASIC PRINCIPLES

|  |  |
|---|---|
| MOVING PARTS | CYCLE |
| UNI-DIRECT-IONAL | MECHAN-ICAL |

[ADDITIONAL THEORY 22]

## FOUR ADVANCED PRINCIPLES

| | |
|---|---|
| **BALANCED ACTION AND REACTION** | **MOMENTUM FROM REST** |
| **MORE HORIZONTAL THAN VERTICAL DISPLACEMENT** | **MOTION FROM ALL POSITIONS** |

[ADDITIONAL THEORY 23]

## LIMITATIONS

A ball rolling a long distance will not have the power to roll up a hill that is equal to the height of the slope.

Therefore, a perpetual motion machine needs a principle that is more efficient than this form of dimensionality. Whether what is required is 2-d or 3-d, it is certain that it requires some kind of D.

**[ADDITIONAL THEORY 24]**

At this point the scientist argues that what is encountered is friction, if not gravity.

However, the perpetual motion designer can argue that what is missing is a very precise configuration.

Since we know dominoes can ascend a staircase, it appears energy can be created, so long as dominoes don't have to be reset.

And I think there is such a case--- and other cases! Again, it depends on a very specific application.

**[ADDITIONAL THEORY 25]**

Nathan Coppedge

## ADVANCED DESIGNS I HAVE DISCOVERED SO FAR:

(1) Horizontal wheel that can perpetuate its momentum [?]

(2) Angularity acting through displacement [?]

(3) Modular units with principle of imbalance [?]

(4) Balance using cheat principle [mostly proven]

[ADDITIONAL THEORY 26]

Remember, that each of the designs is fairly unique. One design might incorporate six or more principles of advantage.

For example, the modular trough leverage uses the following:

1. Begins from rest (no electricity)
2. Moves up and down on its own,
3. Weight versus leverage principle,
4. Uses supporting track,
5. Lever is unbalanced at every point of motion,
6. All parts return to their initial altitudes.

[ADDITIONAL THEORY 27]

*Nathan Coppedge*

# Appendixes

*Nathan Coppedge*

# APPENDIXES

## ADDITIONAL LAWS AND PRINCIPLES

TRINITY OF PERPETUAL MOTION

 I. (a) Total H + V support = 90° at every point, (b) Average altitude is constant,

 II. (a) Momentum without velocity, (b) Unbalanced principle,

 III. (a) Cyclic (b) Mechanical

APPENDIX/

FOR SIMPLICITY'S SAKE

Traditional View of Machines
1. How it's made.
2. What it's for.
3. How it works.

It may be simpler to say that there are two senses of the machine:

1. What it is. Linear thus quantitative, represented by the symbol:

 1. What it does. Typically a spiral

Modified by time to equal a circle:

There is a still simpler case, which is what we have to look out for. The simpler case is simply 'IT IS *SOMETHING*'

Not only does this final view imply commonplace assumptions (the bane of the true philosopher), but under this view one assumes that things of one kind always works one way, which may not after all be the case.

Returning to the three categories we need to prove (reverse of previous): 1. That it works, 2. that it serves a purpose, and 3. That it is not purely theoretical.

So, we begin with theory, we apply purpose, and if the two are flawless, we end up with #1.

It is easier to picture how it might work if we understand that 'what it is' supplements, but does not necessitate what it does (contrary to the conventional view).

We know that if these machines exist, they are physical, and they are machines, and so they do not automatically work unless they are in precisely the correct configuration: they CAN be stopped.

And hopefully started again…

Although starting should involve no more than a relationship of moving parts.

APPENDIX/

APPLIED PRINCIPLES OF PERPETUAL MOTION

**Escher Machine**

Principle: H - V > V(H). Referring to angularity.

Proof: H - V > V(H).

**Modular Trough Leverage Device**

Principle: Supported Leverage

Proof: [1] Individual units have been proven to go up, down, then up again, without using power, and return to their initial altitudes, [2] The repetition of the units requires no additional principle.

**Tilt Motor---or "God's Rolling Pin"**

Principle: Momentum Without Velocity

Proof: [1] A person applying leverage against his or her own weight can lift him or herself smaller distances than the motion of the lever. When this motion is applied to changes in slope, a very small change of distance is necessary for horizontal motion. [2] Because the wheel is horizontal, only slope is required to extend motion. [3] The motion of the levers is greater than the motion to create slope because the rolling cone is tall enough to do so. [4] If resistance is minimal, the result is motion.

### Repeating Leverage

Principle: A counterweight may be operated by leverage. The counter-weight applies constant force, which acts on the lever whether or not it is activated. Thus the lever may be activated repeatedly.

Proof: [1] Slope is possible continuously when upwards and downwards movements are permitted. [2] All that is required for an upwards or downwards movement is a shift in mass; For upwards movement, advantage is required, for downwards movement a disadvantage is required. Thus the system is balanced in force, but unbalanced in momentum. [3] A differing position is acquired by acting upon a mobile ball-weight, which moves because continuous slope is possible.

### Motive Mass Machine

Principle: A domino principle in which the dominoes do not need to be reset.

Proof: [1] A weight in free-fall can move an equal weight horizontally when the second weight is supported on wheels. [2] A chain reaction of this type would gain momentum, because there is a principle of advantage, much like dominoes. [3] In this case, the average altitude remains constant, producing an over-unity principle, because the 'dominoes' 'do not need to be reset': they are re-set, but only by the principle of advantage made possible by the vertical versus horizontal rule, e.g. because part of the weight is unsupported at a given time.

## Mass-Modular Devices: Grav-Buoy 2 Curving Rail Device

Principle: Advantage is had by multiple units, against which is placed minimal resistance.

Proof: [1] In the final design of the Grav-Buoy, the weight of water is cleverly reduced, making the example an exceptional case, [2] In the case of the curving rail, perhaps an absence of friction would make a similar cycle chain-reacting, if not mobile; This has an advantage against some concepts of the Bhaskara Wheel, because the upwardly mobile portion is supported by a track. This is made possible by a flexible chain-link or cable structure. It can be seen for example, that a car on wheels is much easier to push than a car without wheels.

## APPENDIX/ RED LETTER DAYS OF PERPETUAL MOTION

**MARCH 21ST, 2018**
Abdullahi Umar Bessey becomes the first to express real interest in building my designs. He writes, in a private Quora message, that "It's just that I'm tied up with my project otherwise I'd have tried one of yours."

**NOVEMBER 11, 2017**
The 1st or 2nd Scientific Proof of Perpetual motion—Mathematical formulas designed to officially prove the possibility of perpetual motion.

**OCTOBER 2017**
The first possible support from scientists: Ian Switzer writes: "Say, you talk about pulling something up an inclined plane with an equal weight. You're right. This is possible. And not at all a violation of conservation of energy."

**AUGUST 30TH, 2017**
Discovery of the Directive Principle as well as the First Calculus of Perpetual Motion: DP: "It's if the slope changes, and the effect of leverage may vary." PM Calculus 1: "Once you get past the shmoozeballs, it gets tough."

**APRIL 6TH, 2017**
"Brilliant Realization" that there is a loud clunking noise in the ceiling of my college

bathroom, and the lamps along the street near the college seem extra-bright. Could it really be it was Barack Obama who called that one time, asking for rights to perpetual motion?

JULY 12TH, 2016
Day I designed and posted the 1st Fully Proven, a modification of the Successful Over-Unity Experiment 1.

APRIL 5TH, 2016
Day I built the pendulum Type 1, which resembled perpetual motion

MARCH 18, 2016
Day I proved an object could return to the same altitude with momentum.

FEBRUARY 14, 2016
Day I was granted membership to PESWiki.

DECEMBER 15th, 2015
Day I proved a lever could lift its fulcrum with minimal effort.

NOVEMBER 7, 2015
Day I tweeted my most important perpetual motion tweet on Twitter.com (if we ignore Finsrud).

JULY 3RD, 2014
Day I proved an object could roll upwards of its own volition (attempt to build the Escher Machine).

**NOV 10, 2013**
**My First Successful Over-Unity Experiment**

If a device can go up and then down from a position of rest, and all parts return to their original positions, then there is theoretical over- unity. This device meets the criteria.

The diagram shows how to turn the over-unity device into a perpetual motion machine.

**APRIL 26TH, 2009**
**Day I Invented 50 Devices**

These devices contributed significantly to my working tools, but so far as I know none of them have joined my online collection. I cite many flaws.

**APRIL 16TH**
Day in which I have invented Five Devices more than once, first in 2007. One of these was the Coquette.

**APRIL 2ND, 2007**
Day I found evidence that the Tilt Motor could work . With near horizontal slope, leverage can extend slope. Confirmed once, denied twice. Confirmation was with a level.

**FEBRUARY 4TH, 2007**
Day I seemed to prove an unbalanced wheel, the Principled Asymmetry. The device seems to rotate more easily in one direction than another.

*Nathan Coppedge*

**OCTOBER 30TH, 2006**
**Day I invented the Tilt Motor, and time-traveled back to the morning of the same day The Tilt Motor is the ingenious concept of a horizontal wheel, operated by levers.**

## APPENDIX/

### Solving a Flummox

You've seen a design operate significantly at least one time---then the apparatus begins to fail to work. What could cause such a problem? It's called a flummox.

A flummox is a quasi-magical problem that occurs when the powers that be determine that something is in too much doubt to reflect reality. When this happens, even very basic parts cease to work. There is a literal feeling of finding 'gum in the works.' This gum may be invisible, but the powers that be insist that it is there. You can feel them using phony proofs, like 'tape is sticky' and 'it takes a trick,' etc. (However, it does not really take a trick in the absolute sense, at all, unless the principle of combining principles is somehow inauthentic to reality, which it isn't. In fact, combinations make reality much more authentic than it would otherwise be. Complexity is what makes engineering so adequate).

A flummox can traditionally be solved only by the most deliberate work. But another method is to convince a professional manufacturer that the design works as a theory. A still further method is to demonstrate the design in stages---each one of which may serve as proof-of-concept. My Successful Over-Unity Experiment 1 is an example of this.

## BOOKS BY NATHAN COPPEDGE

RELATED BOOKS

    PERPETUAL MOTION PHYSICS
    FOR NON-SKEPTICS
    THE SCIENTIFIC PAPERS

PHILOSOPHY / NON-FICTION
    ARCHE-LOGOS
    BASIC PLATONISM
    THE BOOK OF PARADOXES
    COHERENT LOGIC
    HOW TO WRITE APHORISMS
    INTERMEDIATE INSIGHTS
    METAPHYSICAL SEMANTICS
    THE MODIST MANIFESTO
    THE NINESQUARE NOTEBOOK
    SECR. PRINC. OF IMMORTAL.

ART BOOKS
    HIGH ART
    SUBLIMISM
    HYPER-CUBISM

## THE DIMENSIONAL ENCYCLOPEDIA

*The Dimensional Philosopher's Toolkit*
*The Dimensional Psychologist's Toolkit*

TO BE RELEASED 2015 OR LATER:

*The Dimensional Biologist's Toolkit* (2015)
*The Dimensional Phenomenologist's* (2016)

... ARTIST'S (early-released)
... CRITIC'S (2017)
... EXCEPTIONIST'S (2018)
... UNIVERSALIST'S (2019)

... MATHEMATICIAN'S (2021)
... HISTORIAN'S (2022)
... POLITICS (2023)
... ECONOMICS(2024)

... PHYSICS (2025)
... POETICS(2026)
... TIME-TRAVEL (early-released)
... IMMORTALITY (early-released)

FICTION
    LESSONS OF THE MASTER
    STORY OF MASTER WU
    DRAMATIS PERSONAE
    ONE-PAGE-CLASSICS

*Nathan Coppedge*

# PERPETUAL MOTION MACHINE DESIGNS & THEORY

*Nathan Coppedge*

## BIO

On Nov. 9 - 10, 2013, Nathan finally built a device which he felt proved over-unity. The experiment was followed on July 12th, 2016 with a diagram he felt was fully-proven.

His videos can be found by searching YouTube for "successful over-unity" and "successful perpetual motion experiment".

A more academic-usable video is also present at AcademicRoom.com: "Evidence Against the Classical Model": http://www.academicroom.com/video/evidence-against-classical-model


NOTE FROM THE AUTHOR: I WELCOME REVIEWS OF MY BOOK ON AMAZON, BARNES & NOBLE, AND ELSEWHERE. EVEN YOUR OWN BLOG!

Made in the USA
Coppell, TX
24 October 2021